AF507933

Camp's Botany by the Numbers

A comprehensive study guide in outline form for AP, IB, DE, and college biology classes.

Kenneth R. Camp

Buckethead Publishing

Copyright © [2023] by [Kenneth R. Camp]

All rights reserved.

No portion of this book may be reproduced in any form without written permission from the publisher or author, except as permitted by U.S. copyright law.

CAMP'S BIOLOGY BY THE NUMBERS

Contents

Chapter One

The Evolution of Plants

A) **Algal Ancestors**

1. Terrestrial plants are the plant kingdom's answer to the amphibians' innovations in the animal kingdom. The ancestors of plants are thought to be **green algae (Chlorophytes)**, due to a number of similarities.

2. First and foremost, both types of organisms use most of the same pigments, espially **chlorophylls a and b** to run the **light reactions** of **photosynthesis**. Both taxa use anatomically and biochemically similar **chloroplasts.**

3. In both groups, stacks of pigment filled **thylakoids** are wrapped inside of an **outer chloroplast membrane**. They are submerged in the **stroma**, which is a soup of enzymes that conduct the reactions of the **Calvin cycle.**

4. Both groups use aerobic photosynthesis and split water for electrons and protons to use in the electron chain of the light reactions. Both groups pass **NADPH** and **ATP** onto the **Calvin cycle**.

5. With regard to the Calvin cycle, both groups take in CO_2 as the carbon source that feeds the Calvin cycle, where simple sugars, amino acid precursors, and lipid precursors are made.

6. When the 3-carbon G3P sugars are siphoned off the Calvin Cycle to form **glucose**, both groups tend to convert these monosaccharides into two major polysaccharides, those being **amylose** and **cellulose**.

7. Both plants and green algae store their food reserves as **starch (aka amylose)** inside of **leucoplasts.** Starch has only alpha linkages between the glucose monomers.

8. Plants and green algae create **cellulose cell walls** to serve as protective barriers around their cells, which also protect them from osmotic bursting and **plasmolysis**. Cellulose has alternating alpha and beta linkages.

9. Additionally, with the exception of single-celled and a few colonial varieties, green algae are mostly immobile. None of the filamentous species or **macroalgae** species are capable of independent movement.

10. The diagram below shows commonalities between green algae and their plant descendants.

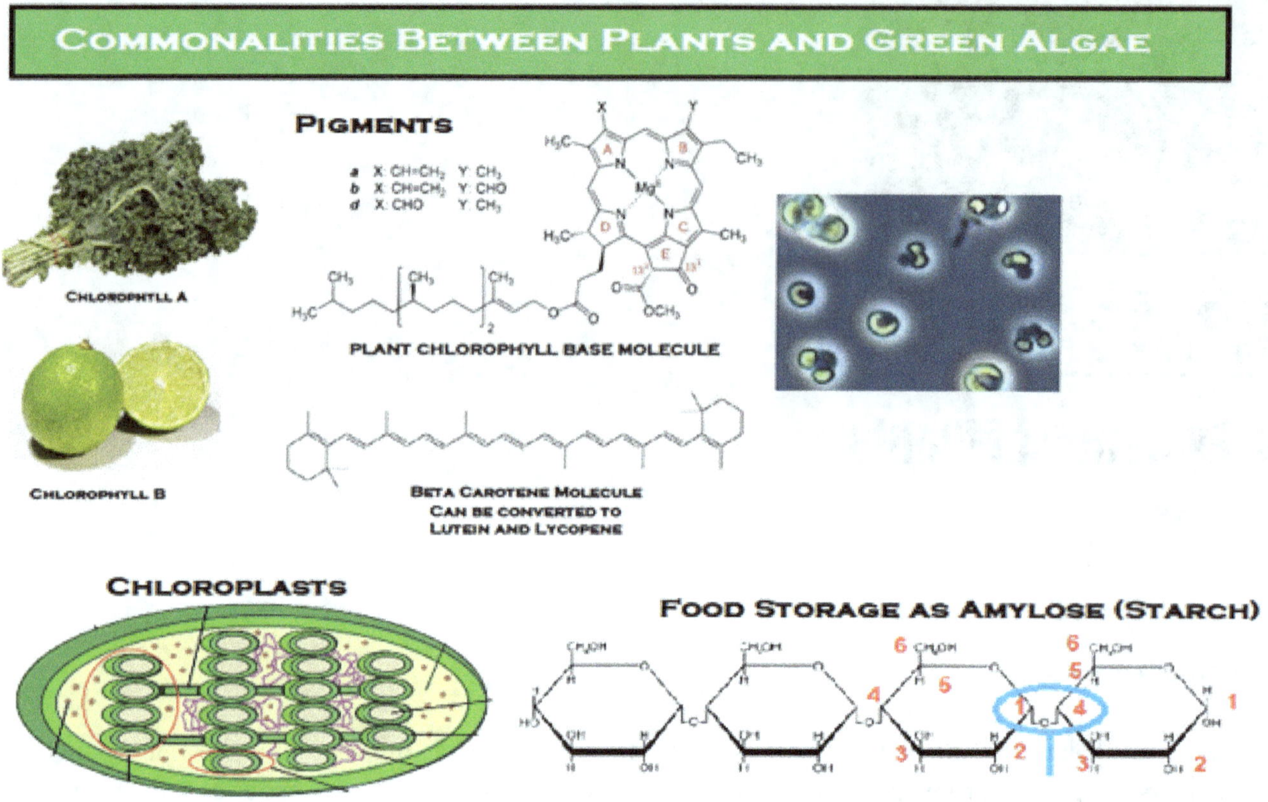

11. Now let's get to the story of WHY green algae needed to evolve into plants.

12. Once upon a time, almost all life lived in the ocean. That meant that there were unlimited occupied **ecological niches** on land. Likewise, many green algae species were (and ARE) found in transient tide pools.

13. Combine an organism that is already semi-adapted to transient drought along with open niches, and eventually it will take adapt further to take advantage of those opportunities.

14. Plants are thought to have moved onto land and diverged from green algae about 400 to 450 million years ago, based on fossil records of the first plants.

15. You might be tempted to think that other phyla of algae are related to plants, but you would be wrong. Algae are **paraphyletic**, meaning that even though many look similar, that doesn't mean they had common ancestry.

16. Therefore, a plant and a green algae are much closer cousins to one another than either is to most of the other algae. While red algae is a distant cousin, according to DNA and biochemical analysis, the same can't be said for the brown algae, in spite of their plant-like apperances. As you can see below, the algae all probably belong in separate kingdoms or clades.

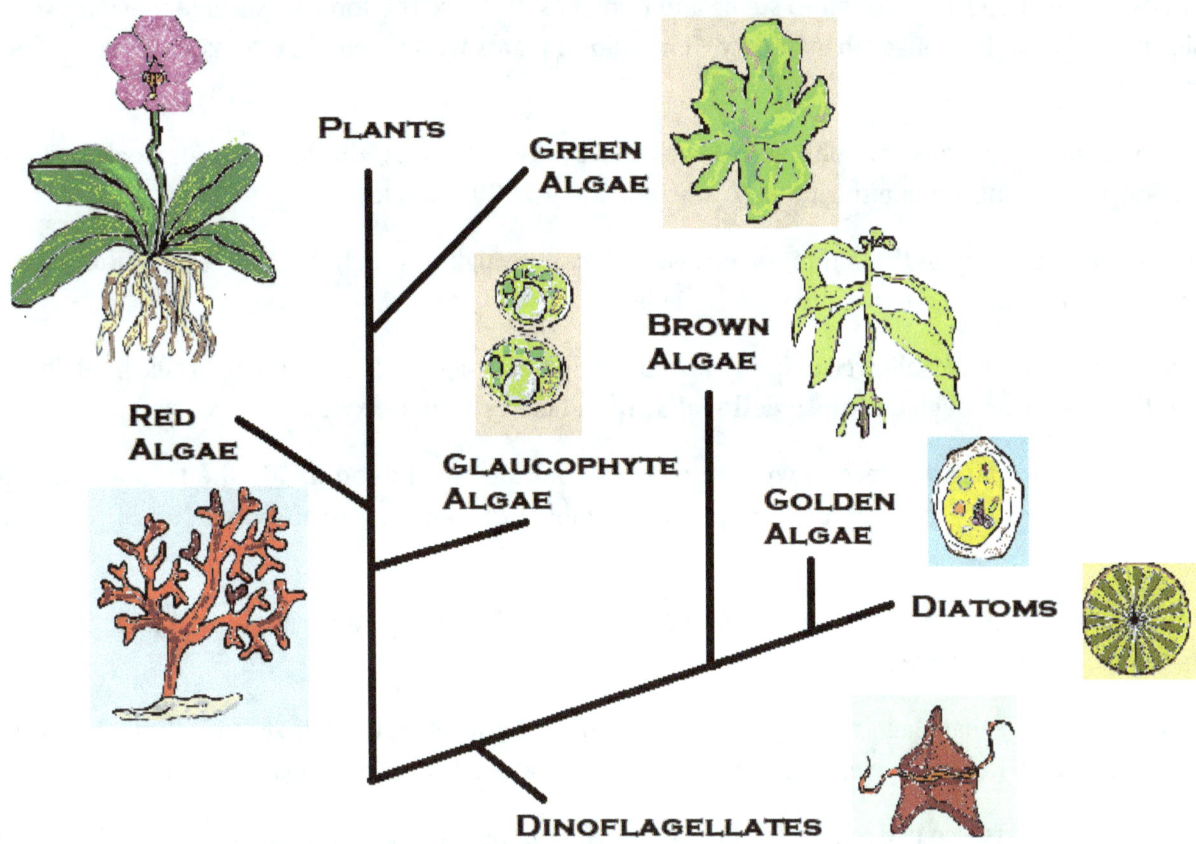

B) Terrestrial Plant Survival Adaptations

1.Like the amphibians also found out, the move onto land allowed organisms to use much larger **niches** and cut down on competition with other aquatic life.

2. However, terrestrial life wasn't exactly like taking candy from a baby. Moving onto land comes with its own set of problems, such as not baking in the sun and shriveling up or figuring out where to get water.

3.Plants had to quickly figure out where they were going to get water, since their tissues were no longer floating in their most important resource. This required the development of **differentiated tissues** and **organs**.

4. **Plants**, except for the primitive seedless bryophytes (mosses and liverworts) all have TRUE **roots, stems, and leaves** with **vasculature** (veins). Mosses, the closest to green algae, are representative of an 'in between' stage.

5. Plant organs are all composed of three types of tissue. These are **dermal, ground,** and **vascular.**

6. **Dermal tissue** is protective and a sealant, **ground tissue** is used in storage, support, and for photosynthesis. However, it is the development of **vascular tissue** that was most critical to adapting to **terrestrial** life.

7. With the exception of the aforementioned bryophytes, all plants have water-conducting tissues called **xylem** and sugar-conducting tissues called **phloem** in all three major organs (leaves, stems, and roots).

8. Mosses are miniature in size, because they don't have the benefit of vascular tissue. Vascular tissue takes advantage of water's adhesion and cohesion to create surface tension and suction.

9. Water will travel up from the soil to stems and branches, because the force of surface tension over microscopic spaces is much stronger than gravity. This allows plants with vascular systems to grow exponentially taller.

10. Compare mosses (no xylem or phloem) to Pacific Redwoods (365 feet tall). Mosses must soak up their water like a sponge, via diffusion, limiting them to a height about 2000 times less.

11. **Algae** don't need any of these things, because diffusion straight through the tissues from the surrounding water serves the same function as the vessels in plants.

12. Another adaptation that allowed plants to be much taller was the development of the rock-hard fiber **lignin** and multiple layers of tough **cellulose cell walls** around the cells in the xylem.

13. Greater support allowed trees to win the competition for sunlight and divide the land into **ecological niches**. Increasing competition created plant specializations and led to the diversity of plants now in existence.

14. However, getting a vascular system and getting taller weren't enough. More alterations had to be made for plants to successfully survive on land.

15. A second resource that plants must have is carbon dioxide. Without CO_2 there is no **Calvin-Benson cycle**. Without the Calvin-Benson cycle, plants can't create organic molecules like sugars and starches.

16. **Terrestrial plants** also had to develop little pin-holes on their leaf surfaces called **stomata** in order to get CO_2 for photosynthesis, since it can't fuse in directly to their tissues from the water.

17. Algae don't need **stomata**, since physics helps them underwater. CO_2 just dissolves and diffuses out of the water and then directly to cells.

18. There are consequences to having pinholes all through your leaves and stems.

19. On the positive side, the negative pressure created by suction of water vapor up through the xylem and out the stomata is a big help to the movement of materials. Water just simply flows upward to lower pressure.

20. On the downside, this means there is a huge problem with water loss from leaves.

21. This is why you can possibly get a few cups of water a day in a survival situation if you just pick a bunch of leaves, seal them in a plastic bag in the sun, and put a drip collector at the bottom.

22. Trees move so much water via **transpiration** through stomata, that this is what creates the weather in rainforest ecosystems. Daily afternoon rain storms result from warm water vapor rising from the trees and condensing.

23. To counter water loss and control it to reasonable levels, plants had to evolve a **waxy cuticle** secreted by the **leaf epidermis** to avoid dessication and water loss.

24. Reproduction also had to change on land. Most algae simply release flagellated **gametes** into the water. Opposite gametes then swim together and **fertilization occurs**.

25. While bryophytes (mosses and allies) and pterophytes (ferns and allies) are still stuck in the past with sperm swimming to eggs in water (rain or flowing creek water), higher plants figured out another way.

26. Most higher plants make use of **pollen granules** to transport their sperm in dry environments. Inside, the sperm stays moist and when it reaches its cone or flower, an accessory **tube cell** grows to the female gamete.

27. All the sperm has to do is swim through the tube cell to the egg, never touching dry air.

28. Speaking of the female egg, higher plant **oocytes** are located **deep** inside the tissues of a **cone** or **fruit**, protected from the environment until sperm find them during **pollination**.

29. Plants have also figured out how to bribe animals to transport their pollen, since their gametes can't swim. Honeybees, butterflies, birds, and bats all do the bidding in exchange for **nectar**.

30. After pollination, plants make use of another strategy. Instead of the **spores** that algae create, plants have also evolved **seeds** that give their offspring a nutrient store. They then use **fruit** to get animals to move offspring.

31. Living and reproducing on land was a tall order that took a long run of evolution to perfect. As we go through this chapter, we will emphasize the evolutionary advances that each group of plants makes over the last.

32. The next section will cover plant reproduction in more detail, since the key to understanding plant evolution lies in understanding their reproductive strategies.

33. The diagram below summarizes some of the evolutionary strategies that plants have made to get by on land. Other than cell walls, none of these structures can be found in green algae.

C) Terrestrial Plant Reproductive Adaptations

1.As previously mentioned, an algae has to do is release its gametes into the water and they swim together.

2. Unlike algae, plants can't exactly throw their **sperm** and **eggs** out into space and expect them to swim together and fertilize. Sperm and eggs can't swim in air.

3. Therefore, plants had to change their reproductive strategies.

4. All plants had to develop reproductive specialized reproductive organs in order to create male and female gametes. All plants have male and/or female **gametangia**, which are the tissues that divide to create gametes.

5. A **male gametangia** is called an **antheridium**. This organ divides, much like testicular tissue in animals, and produces male **gametes** (sperm). In higher plants, these are encapsulated in **pollen grains**.

6. A **female gametangia** is called an **archegonium** and, like ovaries in animals, they divide by meiosis to produce female **gametes** (ova). In most plants, the ova are inside the seeds and the ovary is the fruit or the cone.

7. An easy way to remember this.....**ANTHER**idium sounds like ANTLER and male deer have antlers.

8. Arch**EGG**onium has the word EGG in it, and females lay eggs.

9. Plants use a type of sexual reproduction known as **alternation of generations**, where sexual gamete-producing generations alternate with asexual spore-producing generations.

10. The adults of the **haploid sexual generation** are called male and female **gametophytes**. Male and female gametophytes both have haploid (n) tissues that divide by MITOSIS to make sperm or eggs.

11. Why MITOSIS and not MEIOSIS? Because the tissues of adult plants are ALREADY haploid. They aren't capable of dividing by meiosis, because they are only have one set of chromosomes.

12. Collectively, male and female plants make up the reproductive stages of life known as the **gametophyte generation**.

13. These **gametophytes** are obvious as separate plants in mosses, ferns, and other primitive spore plants, but they are totally hidden inside of cones or flower clusters in higher plants. In flowering plants, they are microscopic.

14. When the male and female gametes meet up, be it through swimming or pollination, they fuse and **fertilization** occurs, yielding a **diploid zygote**. This zygote will grow up into a **diploid asexual generation**.

15. After fertilization, the resulting **zygote** divides by **mitosis** to create an **embryo**, which begins to differentiate and develop tissues and organs. The embryo will eventually mature into an adult **diploid sporophyte**.

16. **Sporophytes** are usually **diploid** (though some plants like bananas and strawberries can be tetraploid or octoploid). When sporophytes reproduce, their **sporangia** divide to produce **haploid spores**.

17. These haploid spores then mature into the **gametophyte generation** once again.

18. A general rationale for why **alternation of generations** came to be, is that in plants like mosses and ferns that still have both generations as separate plants, one is smaller than the other.

19. The **sporophyte stage** is larger and emerges after periods of rain, where the sperm has plenty of opportunity to swim to the egg. Lots of water ensures a large plant can grow too.

20. When drought occurs, the **sporophyte stage**, knowing it may die, releases **spores** which grow into much smaller **gametophyte plants**, which can withstand drought better.

21. When good times come around again, the whole cycle repeats itself again.

22. This bears out a lot of explanation, particularly for plants other than ferns and mosses, because most people don't recognize ANY gametophyte stage in plants.

23. As previously mentioned, plants in the **gametophyte stage** of cone plants and flowering plants are microscopic. They exist only deep inside reproductive parts as dependents.

24. If you look at a pine tree or apple tree, the only thing you can see is the **sporophyte plant**.

25. However, in higher plants, the **gametophyte** lives buried deep down inside the **sporophyte** in the cones or flowers. The larger sporophyte feeds, protects, and cares for the smaller, fragile gametophyte attached to itself.

26. So why does one generation of adult remain attached to the other? Mostly because plants can't move and it creates an enormous inconvenience if they have to try.

27. Primitive plants like ferns and mosses are still dependent on water for **fertilization**. In these plants, the sperm must literally swim to the egg, which is why they're never found anywhere that isn't wet.

28. Higher plants get around this location restriction by casing their sperm inside of **pollen** and making cones or flowering projections that broadcast this pollen.

29. Since most higher plants are **hermaphrodites**, leaving the gametophyte attached to the sporophyte allows for great convenience and effectiveness for the fertilization process.

30. Hermaphroditism allows pollen from two different trees to be dispersed between them, fertilizing eggs that are well protected deep inside the gametophyte that is still attached to the mother sporophyte plant.

31. As we walk through the different taxa of plants, you will notice that reproduction is a central theme. The classification system of plants into phyla, classes, orders relies on differences in reproductive structures.

TRULY TERRIBLE PLANT PUNS

*WHY ARE CONIFERS SUCH HOPELESS ROMANTICS?
  THEY ARE ALWAYS PINING OVER LOST LEAVES.
*HOW DID THE POOR, DESTITUTE BUSH OVERCOME ITS CIRCUMSTANCES AND MAKE SOMETHING OF ITSELF?
  IT ROSE ABOVE IT ALL.
*WHY ARE CACTI RESPECTED AS PLANT PROFESSIONALS?
  THEY ALWAYS LOOK SHARP.

Chapter Two

Non-Flowering Plants: The Bryophytes, Pterophytes, and Gymnosperms

A) **Non-Vascular Plant: The Bryophytes**

1.**Bryophytes** are the most primitive group of plants. They share more commonalities with green algae than any of the more advanced groups of plants. This phylum include the mosses, hornworts, and liverworts.

2. All bryophytes are very small in size, only grow to a few dozen cell layers thick, and remain restricted to moist habitats, since they can't reproduce without a way for their sperm to swim.

3. Bryophytes have NO **vascular tissue**, so they also cannot have true **roots, stems,** or **leaves.**

4. Other than the presence of **stomata** and **rhizoids** (which function a little like roots), there are almost not very many reasons to call them plants, rather than green algae.

5. Like all plants, bryophytes have a life cycle that uses **alternation of generations**.

6. Separate **male** and **female gametophyte plants** are the dominant generation in most species of mosses, liverworts, and hornworts. This is because they are small and stand less risk of dying from lack of water.

7. When there is plenty of rain, the **antheridium** of the **male gametophyte plant** releases **sperm**, which have paired flagella. They look an awful lot like the swimming gametes of green algae.

8. The **sperm** then swim through water droplets to the **female gametophyte plant** and enter the **archegonium**, which is the chamber that contains the egg cells.

9. After **fertilization** occurs, a **diploid zygote** forms and begins undergoing mitosis. After dividing many times, it forms an **embryo** with differentiated cells. The embryo eventually develops into a juvenile **sporophyte plant**.

10. The **sporophyte** then grows directly out of the female gametophyte as a MUCH taller stalk with a spore case at the tip. It remains attached to the female parent for life.

11. Let's say that people were mosses. For argument's sake, we will use Kate Moss as our prototype human female. If Kate Moss had a baby, a stalk the size of a telephone pole would grow out of the top of her head.

12. From there, a spore case the size of a Barcalounger would form. Eventually, when the recliner-sized spore case matured, it would rain a bunch of spores like a shower of BBs. Each BB would then develop into a baby.

13. The **sporophyte plant** has three parts: the **foot** is the base of the plant that absorbs nutrients from the parent gametophyte, the **seta** is the stalk, and the **capsule** forms spores via meiosis and disperses them.

14. Eventually, there will be a dry spell and the **sporophyte** can't get enough water to survive. This triggers cells in the capsule to undergo meiosis to produce haploid spores.

15. As the drought continues, the capsule dries, tightens, bursts and throws out its spores.

16. Since the spores are haploid products of meiosis, they are, by default either male or female. Half the spores carry genes for **female gametophytes**, while the other half contain genes for **male gametophytes**.

17. When these land, they divide by mtitosis to produce baby plants called **protonemae**. These baby plants keep dividing by mitosis and eventually become the adult gametophyte plant again.

18. The diagrams below show the moss life cycle, along with a couple of representative moss species. To be honest, I haven't got the foggiest clue what species they are.....

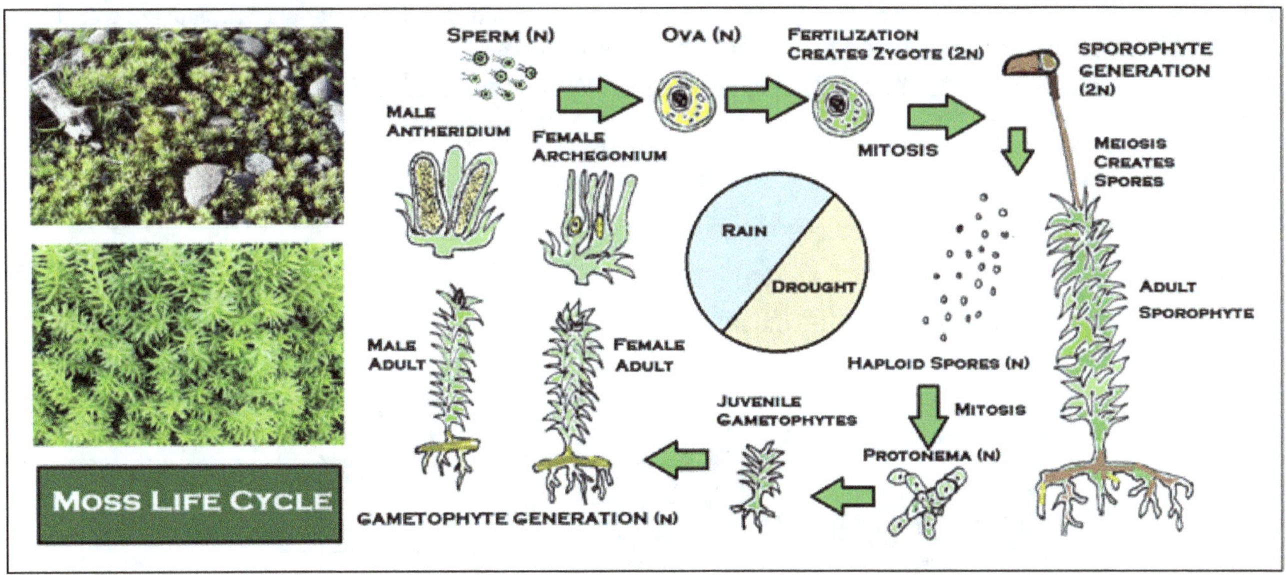

19. Besides mosses, there are two other major groups of **non-vascular plants**.

20. Liverworts and hornworts both have flat, thin sheet like bodies called a **thallus**, rather than fuzzy little leaf-like spikes. They both look kind of like a green tongue sticking out of the ground or glued to a rock.

21. **Liverworts**, unlike mosses, grow a single **gametophyte plant**, rather than separate male and female plants. They are considered to be more evolved because **hermaphroditism** doubles the reproductive capacity of a species.

22. **Male antheridia** and **female archegonia** develop from different clusters of cells on the very same **gametophyte** leaf. Ova develop on the lower edges of the leaf, while sperm develop on the upper parts of the leaf.

23. After sperm from adjoining plants wash onto the top of the leaf of a nearby recipient, the ova are fertilized and pass through the **zygote** and **embryo** stages before a young **sporophyte** emerges out of the top of the female plant.

24. This diploid sporophyte grows on a stalk directly out of the female tissues of the gametophyte.

25. Liverworts can also clone themselves asexually by making little balls of tissue called **gemmae** from areas of undifferentiated tissue. Eventually they loosen from the parent plant and sit down in a little cup on the leaf surface. When rain hits the little cup-like indentations, the gemmae fly off like high bounce balls and start a new plant elsewhere.

26. **Hornworts** are one of the weirdos of the plant world. Their cells only have one giant chloroplast in each cell. This gets them classified separately from **liverworts**, in spite of similar appearances.

28. The table below summarizes the three major groups of bryophyte plants.

ORDER	Sporophyte	Gametophyte	Special Features
Mosses	Stalk grows from egg inside of female gametophyte. Spore case develops at top of stalk.	Typical brush-shaped plants. Separate sexes. Male antheridia and female archegonia	No vascular system, so don't have true leaves, stems, roots. Rhizoids anchor plants in soil.
Liverworts	Flat leathery leaves. Small diploid plant makes spores by meiosis. Can clone by gemmae.	Leathery separate male and female plants grow stalks with antheridia and archegonia.	Rhizoids anchor plants to wet rock surfaces. Budding of gemmae advances over moss.
Hornworts	Horn-like stalk emerges from archegonium, undergoes meiosis and makes spores.	Hermaphrodites, where male and female antheridia and archegonia emerge from same plant.	Single giant chloroplasts in cells. Hermaphroditism is a reproductive advance.

29. Unfortunately for these little obscure plants, they will never be as famous as some of these more noteworth botaniccal bands that hit the bigtime on the music scene.

30. The diagrams below depict representations of hornworts and liverworts.

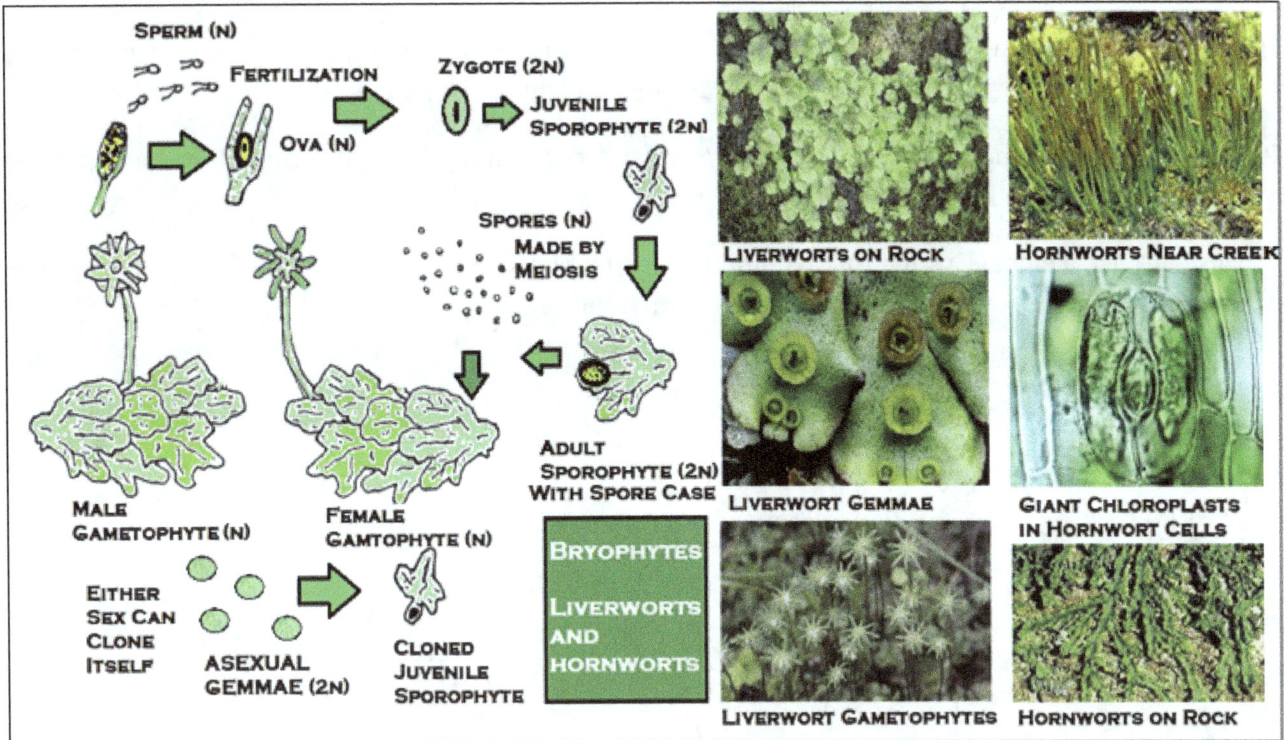

31. Now let's make a quick comparison for why these **non-vascular spore plants** are classified into different orders. While they ALL produce spores instead of seeds and have sperm instead of pollen, there are differences.

B) Seedless Vascular Plants: The Pterophytes

1.The inability of the bryophytes to use anything but **diffusion** for water acquisition put major limits on how much success they would ever have in moving on to land. The **Pterophytes** found a solution to this problem.

2. Water moves by **capillary action** thousands of times faster than by diffusion. Additionally, water molecules can form hydrogen bonds with molecules on the side of a tube by **cohesion**.

3. Over a narrow space, the suction of capillary action is stronger than gravity.

4. A **vascular system** gives plants the ability to absorb water much more efficiently and to grow exponentially taller. **Pterophytes** solved the water transportation problem by developing vessels in their tissues.

5. Therefore, the pterophytes are the most ancient primitive group of plants to develop true roots, stems, and leaves.

6. Pterophytes were the first group of plants to develop specialized vascular tissues. Derived from opposite sides of the same germ-line tissue, **xylem** carries water upward, while **phloem** traffics sugars down to the roots.

7. The presence of **xylem** and **phloem** was a HUGE competitive step up in this group, which allowed its members to get exponentially larger and leave the smaller bryophytes in the dust.

8. **Phloem** and **xylem** allowed these plants to gather water and move sugar MUCH faster and in MUCH greater quantities by **capillary action**, as opposed to the slow pace of **diffusion**.

9. While seed plants eventually evolved and gained a major advantage over pterophytes, there are still species of tree ferns alive today in Asia and Australia that can attain heights of 30 feet.

10. Prehistoric fossils of the extinct club moss *Lepidodenderon* indicated that it formed forests with some individual plants getting to heights of as tall as 120 feet.

11. If that seems like a tall tale, well it isn't. Neither is the first of my stories in the series of bizarre things I've seen in my long career as an educator. Take a study break and read about the first two.....

12. With the ability to grow so large now in hand, pterophytes had to make further adaptations for specialized tissues. Xylem and phloem automatically meant the formation of roots and stems, but leaves would evolve at a later time.

13. While vascular tissues DOES make up the skeleton of leaves, it does not account for the **mesophyll** and dermal tissues that make up the majority of the surface area.

14. Leaves evolved mostly because large plants needs lots of food. The surface area of the stem and branches alone cannot run enough photosynthesis to keep the large volume of plant cells fed.

15. Larger plants needed leaves to increase the surface area of cells able to manufacture sugars by photosynthesis. Green stems and trunks weren't sufficient to feed them.

16. **Leaves** first evolved from extension of stems in pterophytes (the whisk ferns) as the more primitive of two eventual designs.

17. **Microphylls** are the oldest design. These are single extensions of flat photosynthetic tissue off of a single stem. ONLY club mosses still retain microphylls. All other plant types that used them are extinct because they don't work well if you're big.

18. **Megaphylls** are leaves with tissue growth between the webbing of several extensions off of the stem. They are an improved design that give plants much more photosynthetic surface area.

19. Among pterophytes, **whisk ferns** lack any leaves at all, while **club mosses** retain prehistoric microphylls. **Horsetails** superficially appear to have microphylls, but their fossil records suggest that they actually devolved megaphylls.

20. **Ferns**, along with all living species of cone and flowering plants, have megaphylls of some sort.

21. With as much evolutionary progress as pterophytes made, why don't they still rule the world? The answer comes down to reproductive incompetence. They're like a 5-star chef using an E-Z Bake oven to make beef Wellington.

22. All pterophytes still rely on rain or creek water to help transport the sperm from the male **antheridia** to the ova inside of the female **archegonium.** Pterophyte pollen does not exist, nor does any other form of reproduction.

23. Worse, they still rely on the **sporophyte** generation to make free-floating spores that have a very limited amount of time to germinate, should they land on a perfect spot. There is no food storage in a spore.

24. Cone plants and flowering plants solved both problems, so they both evolved right past the pterophytes, leaving most of their members, other than ferns, to relative obscurity in very specialized habitats.

25. Now let's take a look at each individual order of plants within **phylum pterophyta**.

26. As previously mentioned, the most primitive vascular plants are **whisk ferns**, which belong to order **Psilotales**. Whisk ferns lack any leaves at all, and their generation of **sporophyte** plants look like green branches.

27. However, one species native to Australia known as *Tmesiteris* has modified flattened branches that are adapted to function like leaves. However, close examination reveals that they lack most of the tissues found in leaves.

28. Tiny whisk fern **gametophytes** arise from spores produced by the sporophyte in little nodules called **strobili.**

29. The gametophytes are separate sexes, rather than hermaphrodites. Neither gender is photosynthetic, instead living underground amidst similar-looking roots, helped with food by root **mycorrhizae**.

30. Under moist leaf litter, sperm swim to the female **archegonium** when it rains.

31. Whisk ferns are an evolutionary dead-end, since they only have stems and roots. They cannot possibly compete with leafed competitors if they don't have good access to sunlight or help from their root mycorrhizae.

32 At present, there are only 12 known living species of whisk ferns with scattered distribution around the world, presumably remnants of long-since extinct ancestors from many eons ago. The diagram below on the next page shows the whisk-fern life cycle, along with some representative species.

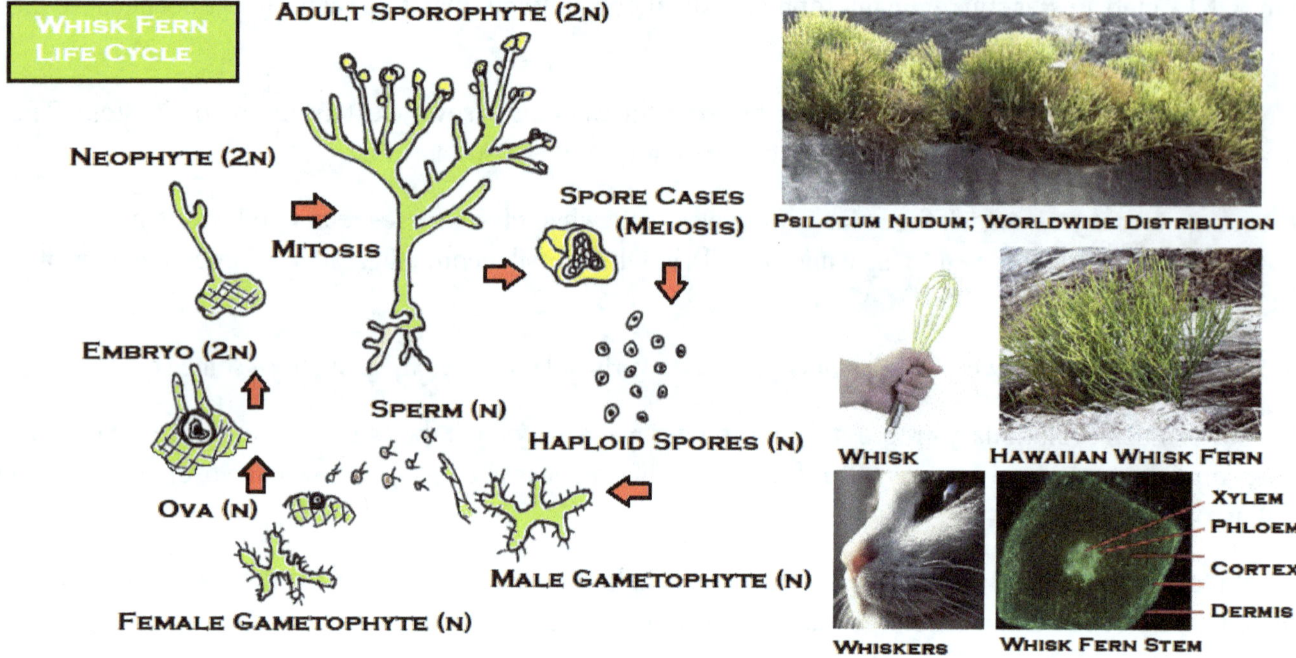

33. Like all other plants, whisk ferns use **alternation of generations** between adult sporophytes and gametophytes. Note the tiny gametophytes in the reproductive generation in the diagram.

34. The next group of pterophytes to evolve were members of the **Order Lycophyta**.

35. At present, there are 335 living species of **club mosses, quillworts,** and **spike mosses** classified as members of this group.

36. Club mosses and spike mosses are actually not true mosses at all, though they superficially seem to be giant mosses with club-like spore cases growing out of the tops. The quillworts could be mistaken for a grass from afar.

37. All three families of plants grow close to the ground as small herbaceous plants.

38. **Lycophytes**, unlike their predecessor whisk ferns, evolved unbranched leaves called **microphylls**, dramatically increasing surface area to do photosynthesis from a straight twig to a fluffy brush.

39. In prehistoric times, club mosses grew to the size of trees and occupied forests as dominant species. You probably know them as the weird looking trees in dinosaur murals.

40. Fossils of scaly bark impressions from their trunks are relatively common in shale rock. Their closest modern relatives are quillwort club mosses, which have retained some of this ability to produce stubby woody growth.

41. Outcompeted by better-designed ferns, cone plants, and flowering plants, today's lycophytes are mostly restricted to niche habitats in wet forest floors, swamps and peat bogs.

42. Reproductively, the lycophytes took a couple of big steps over the design of whisk ferns.

43. Most species of lycophytes are **hermaphroditic**, with spores growing into single gametophytes with **archegonia** and **antheridia** on the same plant. This adaptation doubles the reproductive capacity of each individual plant.

44. However, not all species took the same track of evolution. For instance, **spike mosses** produce separate male and female gametophyte plants, but most of these species are **heterosporous**.

45. In plants with **heterospory**, spores develop differently, depending on their fated gender. Developing male spores are called **microspores**, while female spores are known as **megaspores**.

46. As the name implies, the female spores are larger than the male spores.

47. **Spike mosses** also adopted a strategy very early, which is still used by modern cone plants and flowering plants. The spore cases remain attached to the sporophyte, rather than leaving to become separate plants.

48. **Gametophytes** grow up entirely within the spore cases, rather than dropping off to become a separate life stage like they do in true mosses.

49. Eventually, male gametophytes undergo meiosis and produce sperm cells, and the surrounding spore case cracks open.

50. When it rains or river water splashes onto the spike moss, the sperm swim out of their case and fertilize the archegonia of **female gametophytes** down inside their spore case housings.

51. As the next baby **sporophyte** plant grows, it cracks open the spore case and drops to the ground. Based on how the chromosomes sorted in the meiotic cycle, they will renew the life cycle as male or female gametophytes.

52. The pictures below show the life cycles of a club moss and several representative species.

53. Club mosses can be found growing in moist locations, such as creek banks, low lying wetlands, shady glens, and also under the filtered light of disco balls in places that serve bottles of bub on your birthday. Go shorty.

54. Members of the **horsetail** family have jointed stems and superficially look somewhat like bamboo. Their leaves all protrude at the joints as thin **microphylls.** Fossils suggest they used to have **megaphylls** and devolved them.

55. Apparently there is enough chlorophyll in the stems of to do a sufficient job of photosynthesis.

56. Interesting, since horsetails don't have true wood fiber, yet grow tall, they have developed a strategy of incorporating silica into their cell walls to make them rigid and somewhat inflexible.

57. Pioneers used to call horsetails 'scouring rushes', because they would pick the stems, fray one end, and use them as pot scrubbers to get burnt greasy beans and toasted chipmunk chunks off of their cookware.

58. When horsetail plants reproduce, they develop specializes stems that are tipped with cone-like spore-producing organs called **strobili. Horsetails** are **homosporous,** with male and female spores looking pretty much identical.

59. These spores grow into small separate haploid male and female **gametophyte plants.** Since horsetails are bog plants, the sperm have a relatively easy time swimming from the **antheridium** to a female **archegonium.**

60. Then a new sporophyte develops, grows into another cane, and the next generation begins.

61. The diagram below shows the life cycle of horsetails, along with some representative species.

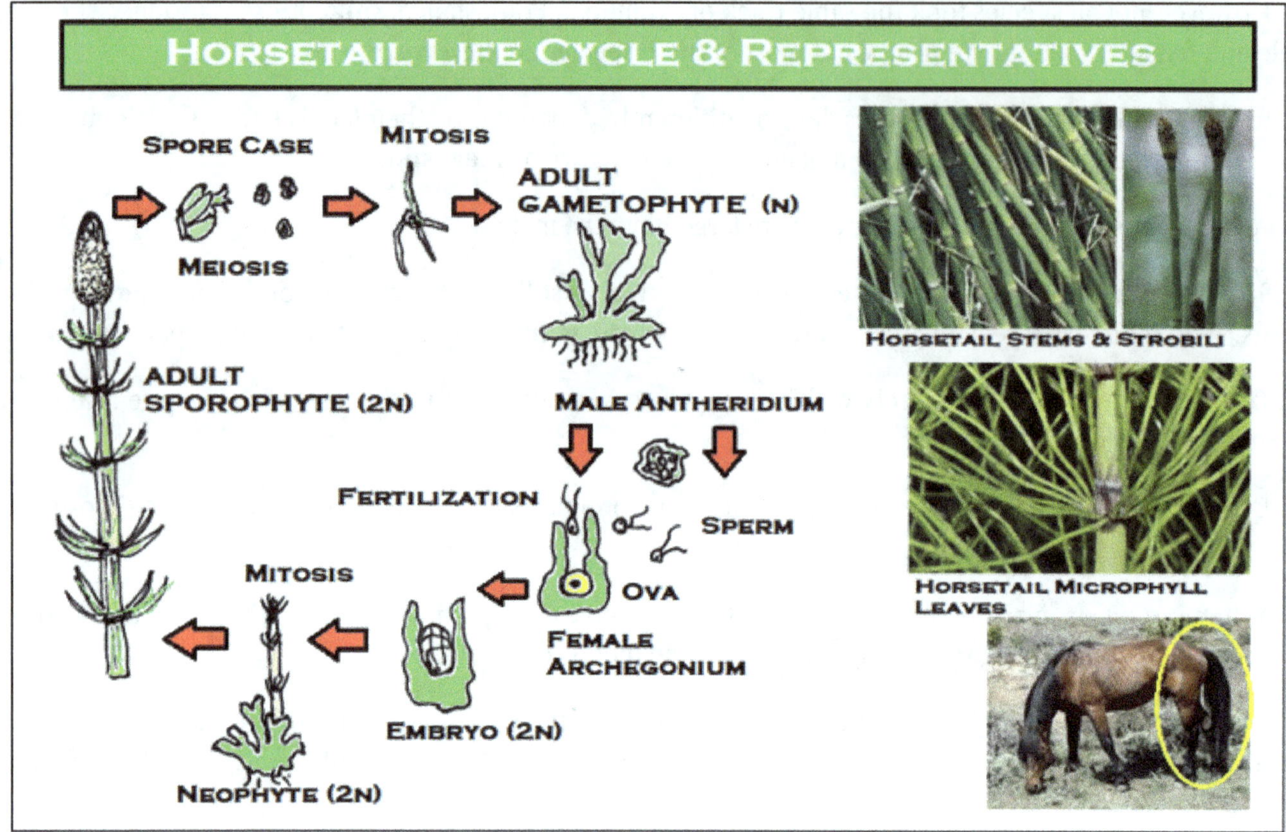

62. In the meantime, enjoy this little blurb about why, under certain circumstances, eating your vegetables is most definitely NOT good for you. Some can kill you, others can embarrass you, while many just taste GROSS. See below.

HEY HOTSHOT, LIKE TO EAT LOTS OF COLLARD GREENS? ENJOY SUFFERING THROUGH GOUT AND KIDNEY STONES

BESIDES THE FACT THAT THEY TASTE NASHTY, RAW LIMA BEANS CONTAIN A COMPOUND CALLED LINAMARIN THAT GENERATES CYANIDE IF EATEN. MAYBE JUST DON'T.

WANT TO CLEAR OUT A BATHROOM WITH A DISGUSTING ODOR WITHOUT EVEN HAVING TO SIT DOWN? EAT SOME ASPARAGUS! THEY TASTE JUST AS BAD!

WANT TO LOOK LIKE A GREAT BIG ORANGE FREAK? EAT TOO MANY CARROTS AND YOU CAN TURN YOUR SKIN A NICE SHADE OF OOMPA-LOOMPA

63. Next time an aggressive vegan tries to preach the green gospel to you, you have some examples to rebut them. Also...remind them that combines kill hundreds of thousands of rabbits and mice when they harvest those vegetables.

64. Ferns are the most advanced group of **seedless vascular plants.** In spite of their ancient origins, they've made a niche existence work, alongside larger more advanced plants. There are more than 10,000 species alive today.

65. The familiar fern **fronds**, that grace the hanging baskets of grandmas everywhere, are the **diploid sporophyte** stage of the plant. **Spore cases** grow on the backs of mature leaves, popping open to drop spores.

66. Fern **gametophytes** are **hermaphroditic** , with **male antherida** and **female archegonia** emerging from different regions on the same heart-shaped plant. In most ferns, the gametophyte plant is about the size of a fingernail.

67. Ferns took a major step up in improving the vascular system, as compared to the lesser evolved members of the pterophytes. Ferns have long underground stems known as **rhizomes** that tightly anchor them to the ground.

68. Roots emerge perpendicularly from the rhizome. In some species, the rhizome may emerge above ground and grow into a fibrous trunk-like stem. Some species of tree ferns native to Polynesia and New Guinea can grow to 30 feet tall.

69. The life cycle of a fern is shown below, followed by pictures of several diverse species.

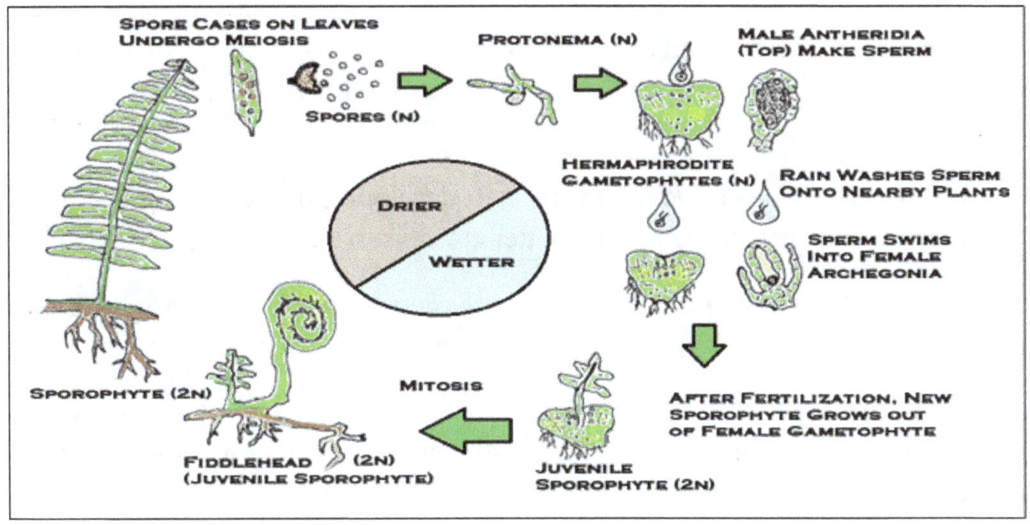

C) The Gymnosperms (Cone-Bearing Seed Plants): Phylum Coniferophyta

1.**Gymnosperms** were the next group of plants on the scene during prehistoric times. They revolutionized the way that plants adapted to land and things have not been the same since.

2. These plants took several quantum leaps forward. They were the first to deploy **pollen grains** to protect their sperm and make water unnecessary for fertilization, the first to invent the **seed**, and the first with true **wood.**

3. **Gymnosperm** means 'naked seed' in Latin. These plants have the name, because their mature seeds are exposed without any surrounding fruit. These either develop directly or are exposed by a mature **cone.**

4. Now let's consider all the reasons that these plants ran right past the bryophytes and pterophytes. Eventually, they would mostly replace them as the dominant group of species in prehistoric forests.

5. **Seeds** ramped up the competition in forests, and represent a MAJOR advance over **spores.**

6. First of all, **seeds** have a protective **seed coat** and a food supply, in the form of a tissue called **endosperm**, which nourishes the developing **embryo** while it waits to sprout. In some plants, this can buy the embryo years of time.

7. Second, the **embryo** of the plant can go ahead and start developing inside the seed. When the neophyte plant hits the ground, it is already much larger and better able to survive.

8. In addition to **seeds**, the **gymnosperms** were the first groups of plants to encase their **sperm** in pollen grains, which allows them to travel independently of water, with no risk of desiccation.

9. Since **gymnosperms** are no longer highly dependent on water to reproduce, it allowed them to spread out and move onto many new niches on land. They provide their own 'pond' inside the protective moisture of the **tube cell** inside the pollen grain.

10. The **seed** is sort of the plant equivalent of the **amniotic egg** in animals. As reptiles put shells on their eggs and ended the amphibian dependency on water to reproduce, so did gymnosperms develop pollen grains.

11. Rather than relying on moisture in the environment, **sperm** inside of **pollen grains** develop alongside a moisture filled **pollen tube cell** that is much larger than the gametes.

12. When **fertilization** time comes, the tube cell continuously grows down toward the egg cell of the female cone. The sperm then enter the tube cell and use it as a canal to swim to the egg.

13. Gymnosperms advanced the idea that began in some of the **pterophytes**. In all gymnosperms, a much larger diploid **sporophyte plant** (such as a pine tree) cares for microscopic **gametophytes** (structures deep inside the cones) for their entire lives.

14. In all higher plants, tiny, **gameophytes** grow directly off of the **sporophyte**. Tiny and helpless, they get all their nutrition from the sporophyte parent and cannot live independently.

15. This dependent relationship ensures a maximal survival rate of the gametes and relieves the gametophyte of the burden of caring for itself, so it can make a ton of gametes.

16. It would not be possible to completely understand the reproductive cycle of higher plants without a microscope. It simply isn't possible to see the parts of the **gametophyte generation** that produce pollen and egg cells.

17. We will cover how these **microspore mother cells** (male gametophyte) and **megaspore mother cells** (female) ultimately end up producing pollen and egg cells shortly.

18. Finally, **gymnosperm plants** were the first to develop **true wood fibers**, in the form of **secondary xylem** layers. This allowed their size to increase dramatically, since conduction of nutrients and support improved.

19. **Xylem tissue**, unlike most plant tissues, is dead at functional maturity. Xylem cells wall themselves off many times with layers of cellulose. Some xylem can have as many as 40 cell walls, each layer adding strength.

20. Since wood fiber is basically millions of xylem cells bundled together, it attains great strength and allows the plant to rise to much greater heights than the fibrous stems of tree ferns, which don't have true wood.

21. The tallest living plants are **gymnosperms**, even today. The Pacific Coast Redwood grows to a height of 380 feet, while the Giant Sequoia trunk can be 90 feet in diameter. Both are **conifers**, which are advanced gymnosperms.

22. For you Star Wars fanatics, who still put their hair in Princess Leia buns and wear Boba Fett helmets as 'adults', you know what sequoias look like. They were where the Ewoks put their treehouses on the Moon of Endor.

23. The wood fiber of **gymnosperms** contains xylem mostly in the form of **tracheids,** which are thin tube-like cells, with a wedge shape and tapered ends. They mostly conduct water in a serpentine fashion through pinholes.

24. Most **gymnosperms** do NOT contain **vessel elements**, which are an improved type of **xylem** tissue that are shaped like flutes stacked end-to-end. Only a group known as the **gnetophytes** have very many of them.

25. **Vessel elements** are more effective at conducting water than **tracheids,** and make up the majority of the wood fibers in flowering plants (**angiosperms**). This is why hardwoods often outcompete pines as forests mature.

26. Tracheids give wood a softer texture, which is why pine wood (soft) is more pliable than hard woods like oak.

27. The diagram below summarizes some of the basic advances we've discussed so far.

28. Now let's focus on the different orders of gymnosperms.

29. Gymnosperm evolution, like most major taxonomic groups, is not completely clear. If you look hard enough, you can find conflicting pieces of evidence between fossil records, DNA sequences, and structural morphology.

30. To the best of our knowledge, the first order of gymnosperms to evolve were the **cycads**. They co-existed with dinosaurs, yet there are still around 300 species still in existence. They have changed very little in the last 100 million plus years.

31. **Cycad sporophytes** look like squat little palm trees or ferns in most cases. Some of them, such as the sago palm, are even given misnomers, since **convergent evolution** causes them to superficially look quite a bit like palm trees.

32. However, **cycads** are about as far away from being palm trees as Amy Schumer is to being funny.

33. The leaves of cycads start developing from the center crown of the trunk called an **apical meristem**. They grow fronds in a rosette pattern. Once the fronds mature, they become very stiff and needle-like.

34. The leaves of many cycads are toxic, because they didn't appreciate being eaten by dinosaurs.

35. Cycads have several carrot-like **primary roots**, with numerous smaller **fibrous roots** branching off of them laterally.

36. Cycads are **dioecious** plants, meaning that there are separate male and female plants. Since **gameto-phytes** remain attached to the leafy **sporophyte** for their entire life, each tree either has male or female organs.

37. Male trees produce pollen from **microspore mother cells** buried deep within male cones. Male cones are technically called **microsporophylls** and are really just clusters of modified leaves bunched tight.

38. Female trees produce ova deep inside the female cones, which are called **megasporophylls**.

39. Cycads are typically pollinated by beetles and weevils that are attracted to compounds made by the plants that seem to mimic insect pheromones. They crawl over cones from both sexes, trying to find mates, spreading pollen grains in their wake.

40. Male and female plants are shown below , along with several representative species of cycads.

41. The next most advanced group of **gymnosperms** retain only a single extant species in their order. **Ginkgos** used to be common in prehistoric times, but today, *Ginkgo biloba* is the only species in existence. It is frequently used in landscaping.

42. **Ginkgos** have woody trunks and root balls that resemble those of flowering plants. Their leaves are fish-tail shaped **megaphylls**. They are unusual among gymnosperms in being **deciduous**, turning bright yellow in the Fall.

43. Reproductively, ginkgos are **dioecious**, with completely separate male and female trees.

44. Male trees produce pollen in pollen cones, or **microsporophylls**, which form from clusters of modified leaves. Male trees are typically used in landscaping, for reasons to be revealed shortly.

45. The ova of female trees develop from **megasporophylls** inside of female ovules that are directly exposed, rather than being inside of a cone. When fertilized, ginkgo embryos develop inside of these, forming hard round **seeds.**

46. There is a fleshy outer layer on the seed called the **sarcotesta** that smells like hot garbage cooking in the sun. Because of this overpowering stench, landscapers avoid planting females.

47. After the seed matures, the sarcotesta splits open on one end to reveal the **sclerotesta**, a hard shell. Wind-blown pollen penetrates a small pinhole opening in this shell.

48. The diagrams below depict pictures of gingkos, including *Ginkgo biloba* males and females.

49. **Conifers** are, by far and away, the most successful of all gymnosperms. There are more than 600 living species. Many of these species are the dominant producers in entire ecosystems, such as taiga and temperate forests.

50. All conifers have true wood, mostly formed from tracheids. They have thick sap and waxy, evergreen needles or scales full of anti-fungal and anti-bacterial compounds, as well as insect repellants, such as turpentine in pines.

51. For many of these species, these **turpenes** and sticky carbohydrates provide cold protection as anti-freeze. Some species of larch live above the Arctic Circle in temperatures, where temperatures can plummet to -50 F.

52. Unlike the **cycads** and **ginkgos**, members of the **Order Conifera** are **monoecious hermaphrodites**. Cones of separate sexes develop on the same tree, making self-fertilization possible in some species.

53. **Male cones**, the **microsporophylls**, develop from soft-modified leaf clusters. The **microspore mother cells** at the base of each cone petal develop into pollen grains. Each grain contains **two sperm cells** and a **tube cell**.

54. **Female cones**, the **megasporophylls**, develop from spiky-modified branches. Female cones are the spiky pine cones that most of us know as grenades from our childhood turf battles.

55. **Ova** develop deep under the petals of each cone in the **megaspore mother cells**. When cones are still green, these ova are buried deep in the cone. Pollen lands on a rosin-filled sticky immature cone, seeking the egg.

56. Male and female cones are shown below, along with the **sporophyte** of a pine tree (remainder of tree).

57. Procreating requires some real persistence because fertilization can take up to a year. The tube cell grows out of the pollen grain, gradually digesting its way toward the ova. When it finally gets there, the sperm swim through it.

58. This arrangement allows plants to reproduce without having to rely on rain or drops from a creek to carry free-swimming sperm. In a sense, just like the amniotic egg put the pond on the inside, so do pollen grains

59. The diagram below shows how **gametophytes** develop from their germ tissues.

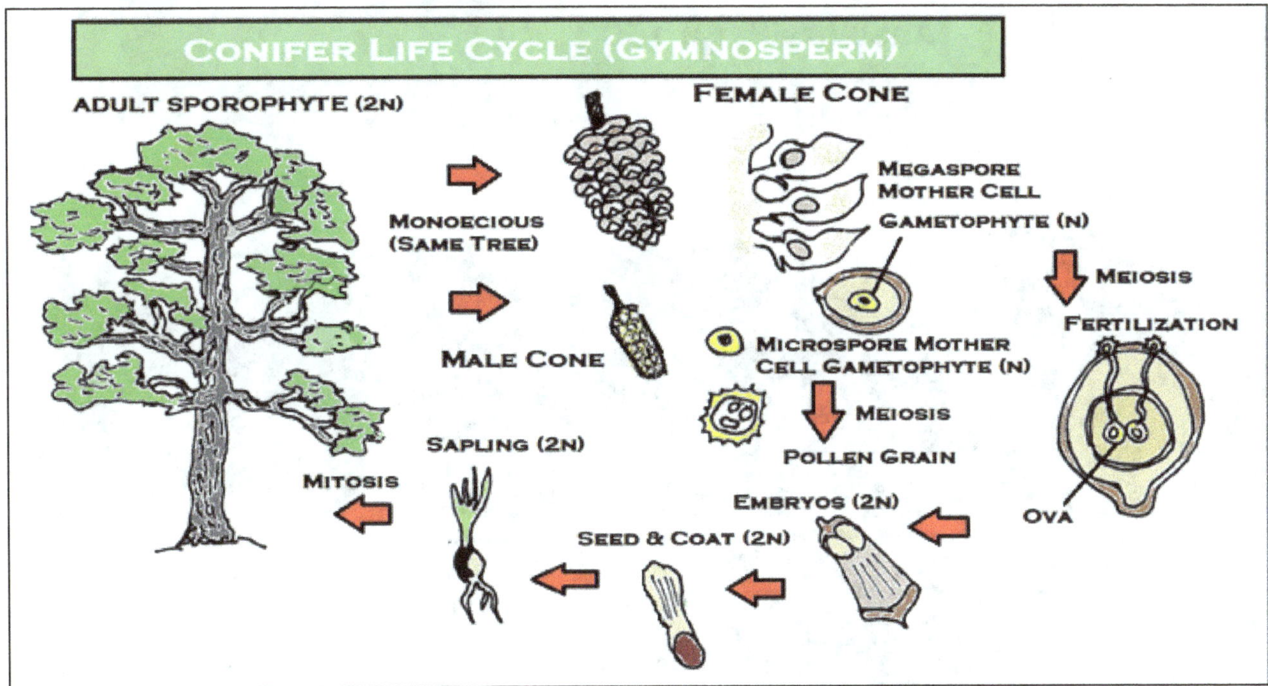

60. The conifers, like the flowering plants, have evolved in numerous directions, insomuch that a layperson might not guess that certain families are close relatives of others.

61. For instance, some members camp the yew family have broad leaves that superficially resemble those of flowering plants, while junipers have tiny needles. Yews have berry-like cones that don't resemble those of pines.

62. The diagram below shows a few representatives of six major families of conifers. But first...enjoy yourself a little study break as you go 'Koo Koo' for shrubbery. Everyone loves landscaping.

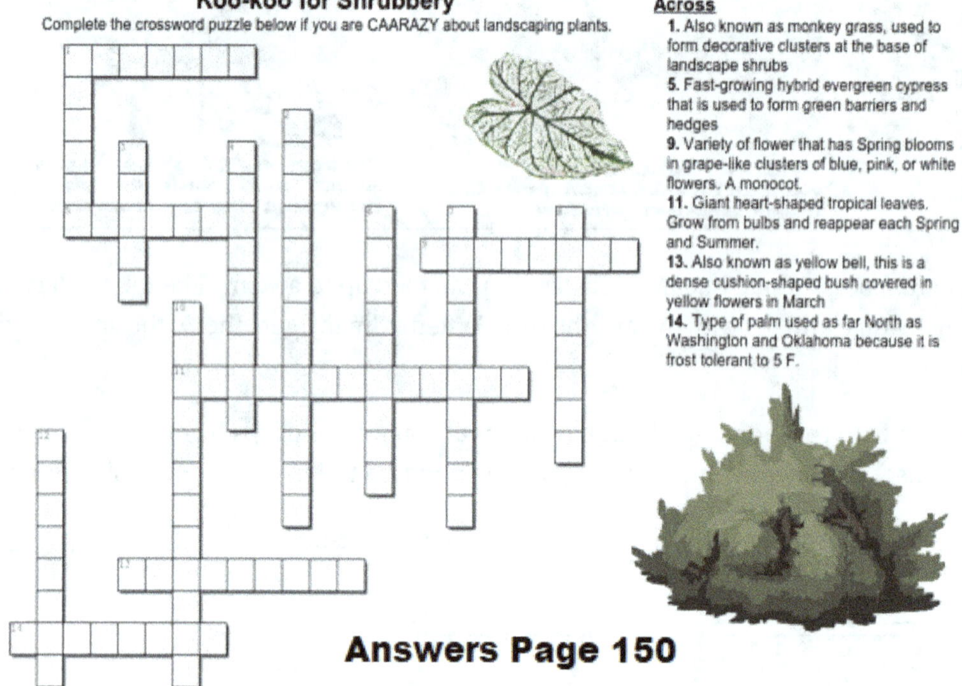

Koo-koo for Shrubbery
Complete the crossword puzzle below if you are CAARAZY about landscaping plants.

Across

1. Also known as monkey grass, used to form decorative clusters at the base of landscape shrubs
5. Fast-growing hybrid evergreen cypress that is used to form green barriers and hedges
9. Variety of flower that has Spring blooms in grape-like clusters of blue, pink, or white flowers. A monocot.
11. Giant heart-shaped tropical leaves. Grow from bulbs and reappear each Spring and Summer.
13. Also known as yellow bell, this is a dense cushion-shaped bush covered in yellow flowers in March
14. Type of palm used as far North as Washington and Oklahoma because it is frost tolerant to 5 F.

Down

1. Staple landscaping bushes of genus Prunus that include English Cherry, Bay, and Mountain
2. Annual flowers planted in the fall that may be gold, yellow, bronze, purple, or white
3. Expensive and elegant species of palm used to line boulevards in South Florida
4. Southern shrub that produces clusters of either blue or pink flowers, depending on soil pH
6. Close relative of invasive Chinese privet people actually buy it. A member of the olive family.
7. Member of apple family. Attractive foliage and bright red berries disguise wickedly sharp thorns
8. Beloved fungus and drought tolerant mutant rose variety that is usually produce magenta or red roses
10. Small trees with smooth bark and sprays of red, pink, or white flowers. Mistakenly cut back each Winter by many.
12. Also known as Narcissus and Jonquil. First flowers to bloom in February.

Answers Page 150

PONDEROSA PINE

JAPANESE RED PINE

JAPANESE LARCH

DOUGLAS FIR

CANADIAN HEMLOCK

WESTERN RED CEDAR

FAMILY PINEACEA
INCLUDES PINES, FIRS, CEDARS, LARCHES, HEMLOCKS,

* NEEDLE OR SCALE-LIKE LEAVES, OFTEN SPIRALING
* DIOECIOUS, WITH SEPARATE MALE POLLEN CONES AND LARGER FEMALE SEED CONES. CLASSIFICATION WITHIN FAMILY IS BASED ON MORPHOLOGY OF CONES.
* TURPENES, PHENOLIC COMPOUNDS, AND STICKY RESIN ALL EMPLOYED AS DEFENSE AGAINST HERBIVORES.
* AROUND 250 KNOWN SPECIES, MANY WITH ECONOMIC VALUE FOR TIMBER, PULP, LANDSCAPING.
* ECOLOGICALLY DOMINANT PRODUCERS IN TAIGA.

Norfolk Island Pine

Kauri Tree

MONKEY PUZZLE

FAMILY ARAUCARIACEAE

INCLUDES NORFOLK ISLAND PINES, KAURI TREES AND MONKEY PUZZLE TREES.

*VARIATION IN LEAF FORMS, BUT ALL ARE EVERGREEN AND GROW IN A SPIRAL.
*LARGE MALE CONES AT BRANCH TIPS.
*POLLEN IS SPHERICAL, LACKS WINGS, RELEASED FROM POLLEN SACS ON CONE.
*FEMALE CONES ARE LARGE AND SCALY AND ONLY PRODUCE ONE SEED PER SCALE.
*MOST ARE GIGANTIC TREES AT MATURITY.
*41 KNOWN SPECIES. ALMOST ALL ARE NATIVE TO SOUTHERN HEMISPHERE.
*BOTH MONOECIOUS AND DIOECIOUS SPECIES IN FAMILY.

Guabisay Podocarpus

Yellow Wood

FAMILY PODOCARPIACEAE

INCLUDES SHRUBS & TREES ALL KNOWN AS 'PODOCARPUS'

*ANCIENT GROUP OF GYMNOSPERMS THAT APPEARS TO HAVE EVOLVED IN GONDWANALAND AND DIVERGED LONG AGO.
*156 KNOWN SPECIES, ALL SOUTHERN HEMISPHERE.
*AXILLARY FEMALE CONES RESEMBLE BERRIES, COVERED WITH A FLESHY MODIFIED BLUE OR GREEN SCALE.
*MALE CONES ARE TERMINAL AND LOOK LIKE CATKINS.
*LEAVES ARE FLATTENED AND RIGID.

F(L) & M(R) Cones of Japanese Umbrella Pine

FAMILY SCIATOPITYACEAE

INCLUDES ONLY ONE MONOTYPIC SPECIES (UMBRELLA PINE)

*DIVERGED LONG AGO FROM YEWS AND CYPRESSES.
*WHORLS OF FLEXIBLE SOFT, NEEDLE-LIKE GROWTHS CALLED CLADODES. THEY ARE STEMS, NOT TRUE LEAVES.
*FEMALE CONES HAVE FLAT WHORLED SCALES, MALE CONES ARE MUCH SMALLER AND SOFTER.
*DATE TO CRETACEOUS PERIOD IN LAURASIA.
*SMALL TO MEDIUM TREES FOUND IN JAPANESE MOUNTAINS.

Pacific Redwood

Giant Sequoia

Common Juniper

FAMILY CUPRESSACEAE

INCLUDES REDWOODS, SEQUOIAS, AND JUNIPERS, CYPRESSES, ARBOVITAE. SCALES, SHORT NEEDLES, PALMATE NEEDLES. TEXTURE ALSO VARIES.
*LEAVES @ 90 DEGREE ANGLE TO NEXT.
*SEED CONES OFTEN SOFT, BRACTED, LEATHERY, SEVERAL OVULES PER SCALE.
*POLLEN CONES ALSO ANGLED RELATIVE TO EACH OTHER, SMALL, SPIRAL SCALES.
*INCLUDES EVERYTHING FROM SHRUBS TO TALLEST, HEAVIEST TREES ON EARTH.
*MOST EVERGREEN, WITH EXCEPTION OF SOME DECIDUOUS CYPRESSES.
*JUNIPER BERRIES (ACTUALLY MODIFIED SCALES) ARE USED TO MAKE GIN.

English Yew

European Yew

Japanese Plum Yew

FAMILY TAXACEAE

INCLUDES THE YEWS AND PLUM YEWS

*ALMOST ALL OF 31 SPECIES ARE FROM NORTHERN HEMISPHERE.
*WAXY RIGID LANCEOLATE LEAVES WITH LARGE STOMATA.
*MALE CONES LESS THAN 5 MM.
*FEMALE CONES SURROUNDED BY RED ARIL. ONLY 1-2 BRACTS MAKE SEEDS.
*ENTIRE PLANT OTHER THAN ARIL ARE FULL OF HIGHLY TOXIC ALKALOIDS.

63. Pines of **Family Pinaceae** are not the only members of the group. Family Pinaceae also includes cedars, hemlocks, larches, hemlocks, and firs. All members of the group are **monoecious** with scale or needle-like leaves.

64. Additionally, all members of Pinaceae have a large difference in size between male and female cones. Once fertilized, seeds fall from the female cones at maturity, using helicopter-like seed coats that blow on the wind.

65. **Family Araucariaceae** includes the familiar Norfolk Island Pine sold at Christmas time, as well as the monkey puzzle tree that grows spike-life triangular leaves up and down its branches to prevent animals from climbing it.

66. Male cones of Araucariaceaen trees grow in **catkins** that look thread-like, while females have gigantic cones. Mature Kauri trees (found in New Zealand and seen often in Lord of the Rings movies) grow to massive heights and have huge canopies.

67. **Family Podocarpaceae** includes shrub-like conifers with soft needles called podocarps. They are often used in landscaping around homes. Like the Araucariaceae, they are native to only the Southern hemisphere.

68. Umbrella pines of **Family Sciadopityaceae** are found only in Japan, and are not true pines. Their cones grow in grape-like clusters. Their plastic-like needles grow in wholes and are broader than those of true pines.

69. **Family Cupressaceae** includes the giants of the conifer world. Pacific Redwoods are the tallest trees on earth, attaining 365 feet, while the Giant Sequoia is the heaviest and widest, growing up to a diameter of 26 feet.

70. While those trees are rare and thousands of years old, more common and familiar members of this group include junipers often seen in shrubbery and the cypress trees that are keystones of Southeastern swamps.

71. Yews belong to **Family Taxacea**. They are small inconspicuous shrubs with soft needles and toxic red berries. Yews make a number of secondary compounds, many toxic. Cancer drugs and gin are derived from yews.

72. Now we move on to the final group of **gymnosperms**, the **Gnetophytes**, whose ranks contain some weird plants that have high curiosity values.

73. The **gnetophytes** of **Class Gnetopsida** are a group of **dioecious** slow-growing desert-dwelling gymnosperms. They have completely separate male and female plants, so both genders must be nearby in order to reproduce. This is not a great strategy for an immobile organism.

74. This was a much larger group of plants in pre-historic times, but the class has dwindled down to just three different classes of oddballs that have only hung on because of their specializations to harsh climates. Hermaphroditic plants have outcompeted them elsewhere, being that most of them don't have to randomly grow near the opposite sex.

75. **Order Welwitschiales** includes a single odd species of plants. **Welwitschias** are freakish plants found only in the barren Namib desert in Africa. As you might imagine, they are endangered and fiercely protected by the government of Namibia.

76. The few thousand plants in existence have just two long leaves that grow continuously for life. It is a long life, indeed, since individual plants can live to be more than 1,000 years old.

77. Clusters of seed-bearing cones develop slowly from the center of the female plant.

78. It's at this point, that the author must make a confession. It's true, I nearly killed one of the said few thousand plants in existence, and a member of one of the ONLY collections in the U.S.A.

79. Once while on tour of the University of Georgia greenhouses, I was handed a potted Welwischia to show my students....one of the few that actually exist outside of the country of Namibia.

80. I lost of my grip and dropped the pot before making a miraculous last-second catch at ankle level. With the curator's back turned, I looked up to see several wide-eyed students. They never told on me either....so it 'never happened'.

81. **Order Ephedraceae** includes a few dozen species of mostly scrub-land and desert shrubs, vines, and small trees. At first look, you might not guess that many such species are actually gymnosperms, and not flowering plants.

82. They have green photosynthetic stems, scale-like leaves, and **strobili** at the tips of branches that can superficially resemble flowers. Most species are **dioecious**, with separate sex plants.

83. Some species of **ephedras** produce stimulant alkaloids similar in structure to caffeine. Mormon tea shrub produces **ephedrine**, which was used to prepare tea for tee-totalers.

84. This compound can be further synthesized into **pseudoephedrine** used in cold medicines.

85. In case you wondered, yes, you CAN draw a line back from a meth lab to an ephedra plant, though ephedrines are now mostly made synthetically, rather than gathered from plants.

86. But, yes, a dilapidated single-wide trailer (with a toilet flower pot on the front lawn), at some point, HAS exploded, indirectly, because of the compounds made by these plants.

87. **Gnetums** of **Family Gnetophyta** are evergreen trees, vines, and shrubs. Many of these species are tropical and you wouldn't guess that they're related to the aforementioned groups.

88. Most gnetums bear a striking resemblance to regular angiosperm plants, as some have broad leaves and produce female cones that resemble seeds or small fruits. However, the male **staminate cones** give them away.

89. Additionally, the wood fiber has mostly **tracheids** rather than **vessel elements**, like most gymnosperms. They also have a lower photosynthetic rate than flowering plants, because they have fewer chloroplasts in their cells.

90. Examples of **gnetophytes** are pictured below.

WELWITSCHIA

EPHEDRA PLANT

GNETUM SHRUB

Chapter Three

Flowering Plants: The Angiosperms

A) **Characteristics of Phylum Anthophyta (The Angiosperms)**

1.**Angiosperms** are by far and away the most common plants on earth today. Flowers, fruit, and improved vascular systems have given them an overwhelming advantage over all other types of plants.

2. It is estimated that around 95% of all living plant species are angiosperms.

3. With the major exception of timber and paper and a handful of pharmaceutical drugs, almost all plant products that people depend upon come from angiosperm plants.

4. There are several good reasons this is so, and they all have to do with evolutionary advantages in competition for resources and habitat niches. Cone plants and lower plants simply couldn't hang with flowering plants and lost most niches.

5. **Angiosperms** show dramatic **divergent evolution**, even among closely related groups.

6. Flowering plants have been able to evolve a plethora of dramatically different morphologies. From cacti to grape vines to oak trees, angiosperms have adapted to use of every available terrestrial niche.

7. One major advance that occurred among the **angiosperms** is more efficient conducting tissues. With a better vascular system, flowering plants often have an advantage over cone plants competing for a space.

8. **Vessel elements**, which are pipe-like and conduct water up and down, as well as sideways, are dominant in the **xylem** of flowering plants. They conduct more water at greater speed, than tapered **tracheids**.

9. All things being equal in a warm, wet climate, an angiosperm tree will almost always outcompete a gymnosperm tree for water and nutrients. This is part of the reason temperate forests show **ecological succession**.

10. Many temperate forests are dominated by pines in their early stages. As time goes by, hardwood seedlings gain ground, and outcompete the pines. By the time it reaches a **climax community**, oaks and hickories dominate.

11. **Angiosperms** also got ahead in the competition for terrestrial niches by improving their reproductive capacity.

12. **Flowers** and **fruits** both take advantage of the mobility and behaviors of animals to help them reproduce.

13. While some **angiosperms** (particularly monocots like grasses) are still wind-pollinated, and have **incomplete flowers** with reduced petals or no petals, most use **pollinators** to dramatically improve reproduction.

14. By offering a bribe of **nectar** to pollinators like birds, bees, beetles, flies, and bats, they dust the animal with pollen as they drink. From there, the animal transfers pollen to the **pistil** of the next flower.

15. The pictures below show several plants with different pollinators.

SWALLOWTAIL ON HIBISCUS   HAWK MOTH ON FUSCHIA   HUMMINGBIRD ON TRUMPET VINE   BAT ON CACTUS

16. **Petals** evolved in conjunction to serve the same purpose that a gigantic 'Arby's' sign does to tell diners to come get some roast beef and curly fries. They are advertisements for food.

17. Instead of paying for their food in cash and coupons, the fee for a meal is to carry pollen to the next plant.

18. With that said, we must now examine the anatomy of flowers more closely before we continue.

19. The table and diagram below illustrate the anatomy of a flower.

20. Keep in mind that the diagrams illustrate a **perfect flower**, in that the flower is a **hermaphrodite** with parts of both genders on the flower. **Imperfect flowers** have only one set of gender organs.

21. Plants with both gender organs in the same flower are **monoecious**. Examples of such plants are lilies and roses.

22. Hermaphroditic plants with different male and female flowers on the same planta re **partially monoecious**. Examples of such plants are bananas, squashes, pumpkins, and gourds.

23. **Dioecious** plants have two genders on completely different plants. They are relatively rare among flowering plants, but still exist. Some examples are pomegranates, persimmons, muscadines & kiwi fruit.

24. The diagram below illustrates examples of dioecious, partially monoecious, and monoecious species.

DIOECIOUS PLANTS: Lily & Hibiscus. Hermaphrodite Flowers.

PARTIALLY MONOECIOUS PLANTS: Squash & Bananas.....Separate M & F Flowers on Same Plant.

MONOECIOUS PLANTS: Pomegrantes have separate M & F Flowers on Different Trees

25. Now let's examine floral anatomy. We will cover the example of a **complete dioecious flower**. There are many differences in flower design, but the general principles of the function of individual parts is the same.

26. In dioecious plants, individual flowers are **hermaphroditic**, with both male and female parts.

27. Flowers come from **terminal buds** or **lateral buds**, depending on the species. The **calyx** is the first part of the flower that is visible as it is developing. As it opens, these modified leaves become the **petals** and **sepals**.

28. The petals and sepals are really just parts of the **diploid sporophyte** that have been modified to advertise nectar availability for pollinators, and to protect the young developing **gametophytes**.

29. Collectively, the female organs of the plant are known as the **pistil**.

30. The top of the pistil is a sticky pad called the **stigma** that collects pollen off the bodies of pollinators. The **style** is a long tunnel that allows the **pollen tube** to grow down into the **ovules** after pollination.

31. At the bottom, the **ovary**, which is fated to become the future fruit, generates and protects the **ovules**, which will be fertilized when sperm swim through the **pollen tube**. After fertilization, they become **seeds**.

32. The actual **female gametophytes** of flowering plants are cells deep within the **ovary** called **megaspore mother cells** that divide by meiosis to create the ovules. We will cover this in more detail later in the chapter.

33. The male organs of the flower, collectively are known as **stamens**.

34. The stamens make use of modified stems called **filaments** to support the **anthers**, which generate **pollen**.

35. The anthers need to be above the surface, because they are essentially dusting wands that insects, birds, or bats will bump into as they dive into the flower to drink the nectar at the base of the petals.

36. The actual **male gametophytes** are deep within the **anthers**, and are called **microspore mother cells**. They divide by meiosis to create grains of **pollen**.

CIRCA 2005

IT WAS JUST ANOTHER ROUTINE DAY OF HIGH SCHOOL. THAT WAS ABOUT TO CHANGE. FROM SEEMINGLY OUT OF NOWHERE, A RAGING BALL OF PORCINE FURY ANNOUNCED ITS ARRIVAL WITH EAR-SPLITTING SQUEALS, AS ALL CALCULUS, CHEMISTRY, BRITISH LITERATURE, AND BIOLOGY CLASSES CAME TO A SUDDEN SCREECHING HALT ON THE BACK HALLWAY. CORNY THE PIG SPRINTED THROUGH EACH AND EVERY CLASSROOM, FOLLOWED CLOSELY BY THE AG TEACHER, SEVERAL STUDENTS, AND THE PRINCIPAL. NO ONE WAS ABLE TO CORRAL CORNY AS HE BROKE TACKLES LIKE A CRIMSON TIDE RUNNING BACK. CORNY FOUND HIS WAY OUT THE BACK DOOR AND ESCAPED INTO THE WOODS, REMAINING A FUGITIVE FOR SOME TIME, IN SPITE OF A SCHOOLWIDE APB (ALL PORK BULLETIN).

The Life of an Educator.....

The Dumbest Moments of my Esteemed Career

Episode 3

Sooie Pig!

37. The diagram below shows a lily, which is a pretty generic dioecious plant. The functions of each flower part follows.

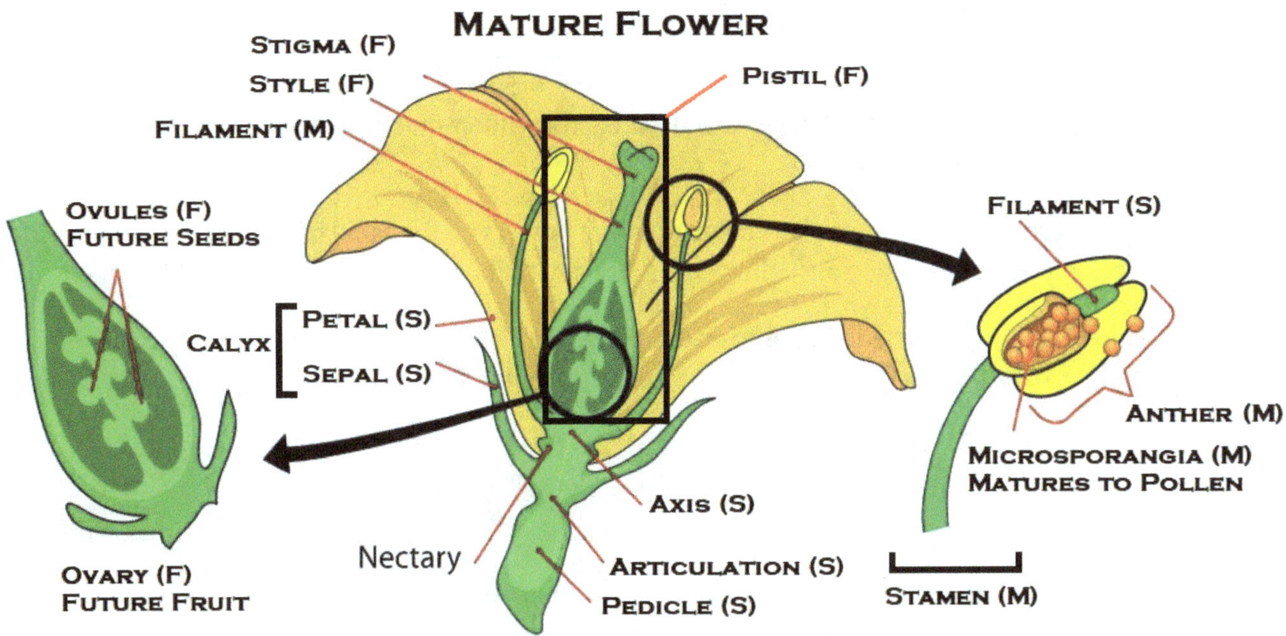

ORGAN	DESCRIPTION	GENERATION
Sepals (Calyx collectively)	The leaves that open around the flower bud. In some monocots, these are colored and help the petals to attract the pollinator. **FUTURE FATE: Fruit scars**	Sporophyte
Petals (Corolla collectively)	The modified colored leaves that open around the male and female flower parts. Their coloration attracts pollinators **FUTURE FATE: Fruit scars**	Sporophyte
Pistil	**Collective term for the female organs in the center of the flower.** Includes ovary, style, and stigma.	Female gametophyte
Ovary	Female flower organ in the center of the flower. Usually bulb-shaped and notably swollen. Megaspore mother cells inside of it divide by meiosis to produce megaspores, which give rise to ovules. **FUTURE FATE: Fruit body**	Female gametophyte
Ovule	Haploid female gametes that are the result of meiosis. These are fertilized by the pollen when the pollen tube grows down into the flower. **FUTURE FATE: Seeds**	Female gametophyte
Style	Long tube-like projection that leads to the ovary. The pollen tube must grow through this to reach the ovules. This allows the time the ovules need to mature. **FUTURE FATE: Fruit scar**	Female gametophyte
Stigma	Sticky pad that secretes nectar on top of the pistil. Attracts pollinators and also traps pollen in its sticky secretions.	Female gametophyte
Stamen	**Collective term for the male organs on the periphery of the flower. Unlike the pistil, there are several of these on any given flower.**	Male gametophyte
Filament	Stalk-like projection that holds pollen-producing anthers up for pollen distribution. **FUTURE FATE: None**	Male gametophyte
Anther	Male gametophyte contains microspore mother cells that divide by meiosis to produce microspores that give rise to pollen grains. **FUTURE FATE: None**	Male gametophyte
Pollen	Two haploid male gametophytes, which are contained inside of a pollen coat, along with a tube cell. **FUTURE FATE: Fertilization of the ovule**	Male gametophyte

38. There are dozens, if not hundreds, of designs of flower across different taxa in nature, but the fundamental principles of each organ remain the same.

39. While male and female gametophytes may be on different flowers on the same plant (partially monoecious) or different plants altogether (monoecious), the cellular biology of the gametophyte remains the same.

40. Some dioecious plants also produce **compound flowers** that may look like a single **influorescence**, but the same flower may have dozens or hundreds of stamens and multiple pistils.

41. Regardless of anatomical layout, the basic parts and functions are similar to what is depicted in the lily diagram on the following page. Until we get to the next page, let's fill up this empty white space with something random.

42. As you can imagine, writing a textbook can be a very tedious process, and an author's mind is going to wander considerable distances away from the main point. Right now, I need a GPS to figure out where I'm headed.

43. Unlike most other authors that prudently edit their insane ramblings out of their final copy, that's not me. You, dear reader, are along for the ride, because 'professionalism' is just a word to me.

44. Today my sister had to go to court because some woman claimed that she gave her whiplash five years ago after this genius of a con artist pulled around my sister's truck and immediately slammed on her brakes.

45. She works as an 'actress' part time and makes the rest of her money by defrauding people. Oh wait…she does that full time anyway, because she is an 'actress'. Sadly, as dumb as this is, it is a typical lawsuit.

46. That got me thinking, just how dumb can lawsuits get? So I decided that this would be a good space-filling topic.

45. Now that you've familiarized yourself with the working parts of a flower, you need to understand what's going on inside the parts of the flower. The inner workings of plant reproduction aren't as obvious as they seem.

46. Remember that plants reproduce by **alternation of generations**. NO exceptions. Just because you can't SEE separate male and female gametophyte plants doesn't mean they aren't there.

47. Where are all the smaller sized **gametophyte plants** if this is the case?

48. In a disappearing act that would make Harry Houdini proud, the plant shrank them to microscopic size and hid them down inside the deepest parts of the flowers. This provides protection and insurance of a next generation.

49. The much, much larger **diploid** (usually) **sporophyte plant** acts as a nourishing protector of the **gametophyte**.

50. Pretty much any plant part you can actually see is part of the sporophyte, be it an apple tree, blades of grass, or vegetable plants in your garden. The gametophytes are so microscopic that they aren't visible.

51. This makes sense from an evolutionary standpoint, because trees and other flowering plants, unlike mosses, have vascular tissue. This gives them a distinct survival advantage here too.

52. It doesn't make sense for plants to waste time, energy and resources producing huge numbers of tiny gametophytes that mostly don't survive competing against larger sporophytes of other species.

53. Since the larger sporophytes have adapted to deal with rain or drought, it makes more sense for them just to take care of the gametophytes, rather than releasing them to become separate plants.

54. This also gives the plant the ability to continuously reproduce, in many species, rather than having to take breaks for weather between the diploid and haploid generations.

55. Continuous reproduction has the positive consequence of constant meiosis and more genetic variety.

56. Now let's focus on the processes that create gametophytes in both male and female plant organs.

57. Inside the **anthers** of male plants are cells known as **microsporocytes** or **microspore mother cells**.

58. These correspond to the spermatogonia inside the testes of male animals. These tissues divide by meiosis to produce **microspores**. These cells are taken care of by another layer of cells called the **tapetum**.

59. Much like the Sertoli cells nourish developing sperm in the testes, the tapetal cells fill the developing microspores with nutrients that get stored in vacuoles.

60. From there, the **haploid microspores** undergo mitosis, but the cell does not divide evenly.

61. One cell retains 95% of the cytoplasm and is filled with a giant nutrient vacuole. This nucleus will become a **vegetative nucleus** as part of the future **pollen tube cell.**

62. The other new nucleus is barely much more than a nucleus covered with a cell membrane. This nucleus will be fated to become the two **sperm** inside of a pollen grain.

63. As the pollen grain matures and moves closer to existing the anther, the vacuole breaks up into many different smaller vesicles. The outside of the microspore hardens into the coat of the **pollen grain.**

64. The diagram below summarizes the basic process of pollen formation up to the point of pollination.

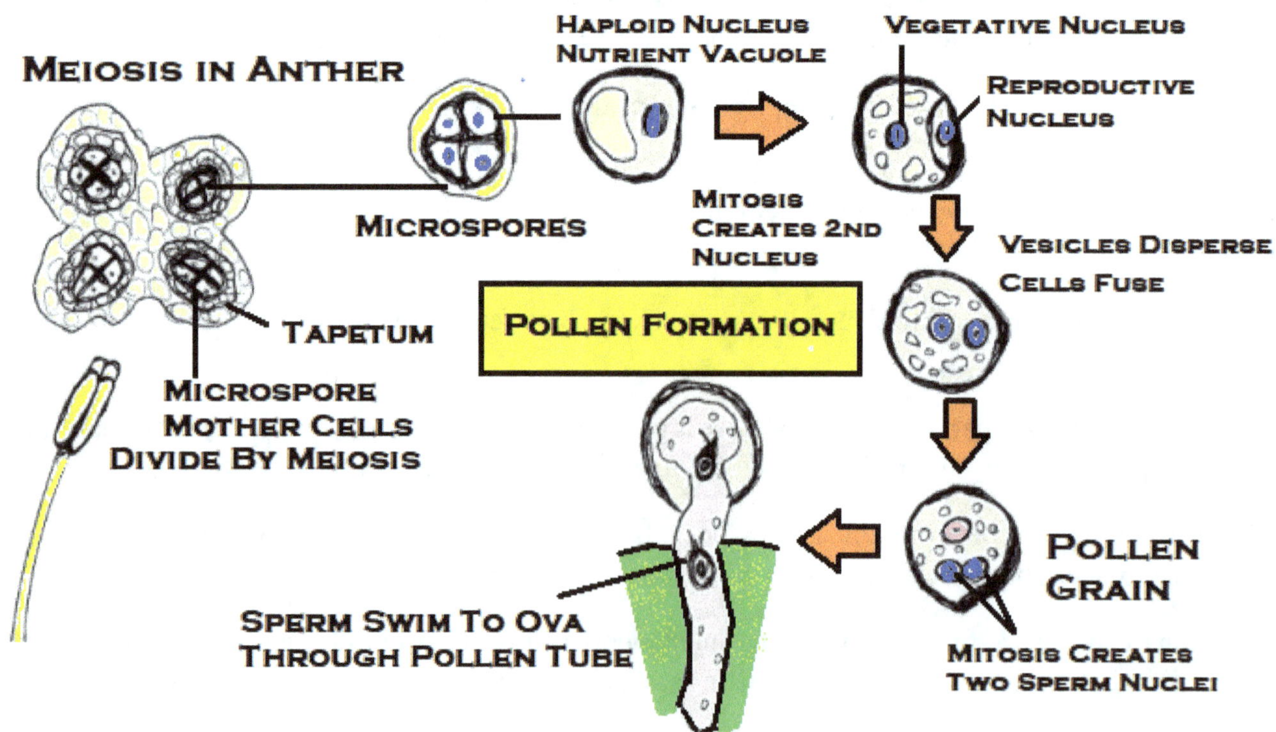

65. Now let's consider what happens in the female **pistil**. The idea is similar, but the process differs somewhat.

66. Deep inside the ovary at the base of the pistil, the **female gametophytes** form from the division of cells known as **megaspore mother cells**. These diploid cells undergo **meiosis** to form four haploid **megaspores.**

67. Similar to what happens in egg development in the ovaries of animals, only one of the four cells will survive to develop into an egg. This lone cell hogs all the nutrient vacuoles and starves out the other megaspores.

68. These cells are wrapped in two layers of tough protective tissue called an **integuments**. A pinhole in the center of these tissues, known as the **micropyle** allows access for a future pollen tube to access the ovule.

69. The **integuments** will become the tough strings that hold the seed coat to the fruit after pollination. To see a great example of an integument, cut a hot pepper in half. The stringy white stuff is the integument.

70. At first there all four megaspores are alive inside the integument, but as the other three unused megaspores kick off and die, development continues to get the surviving megaspore into position to be a functional egg.

71. This process can vary widely among different classifications of plants, especially those that are **polyploid** instead of **diploid**. In fact, this development process has so many variations, we will keep it simple here.

72. We will only consider the most common type of developmental scheme seen in the largest number of flowering plants. The lily is a good type specimen for this system of reproduction.

73. In lilies, the surviving **megaspore** then undergoes mitosis three times in a row. It is the size of cells after these divisions, and the fate of daughter cells after mitosis that can vary so much between different families of plants.

74. Regardless, there are a couple of unusual things about these mitotic divisions. First, the nuclei are **haploid**. Second, the cytoplasm doesn't divide evenly, because the cells are all fated for different functions.

75. These modifications all have to do with functional needs, once the megaspore becomes an **ovule** and a **seed**. We will consider each division individually.

76. In lilies, following the haploid megaspore divides to create two nuclei, then four. At this point, one haploid nucleus remains in the center of the cell, while the other three migrate toward the **micropyle**.

77. Since the center nucleus in the three cells clustered under the **micropyle** is in the best physical position to be **pollinated**, it is selected as the future egg cell. The other two cells on the side become **synergid cells.**

78. The **synergid cells** are like 'wing women' who keep trying to fix up their single friends with blind dates. They effectively scream and holler at the pollen grain by sending out biochemical signals to attract the pollen tube.

79. While we are the topic, if there are single male readers here, the author would encourage you to NEVER allow a group of well-meaning female 'friends' to try to 'set you up' on a blind date. Good things RARELY happen. See below.

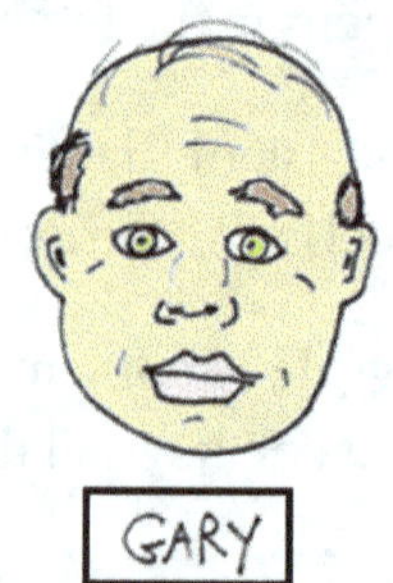

80. There are a few common outcomes of such misguided goodwill gestures. There is an approximately 76% chance one or both of you will feel insulted that your 'friends' would place you in the same league with each other.

82. If you escape this fate, there is still a 15% chance that the prospect in question will declare 'love at first sight', pick out 'your song' and begin to name future children before your leftovers have cooled and dessert arrives.

83. There is also a 5% chance that one (or both of you) will bring up a political stance, religious belief, or gender role conviction that will instantly enrage the other party, who will begin to argue in an attempt to 'fix' you.

84. At this point, go ahead and talk about your past relationships, because it's over anyway. In either situation, excuse yourself, escape by climbing out of the bathroom window and change your phone number.

85. There is also a small, yet not insignificant chance, that your date has thought about what it might be like to be a serial killer. If you doubt this, check out the 1979 episode of 'The Dating Game' with Rodney Alcala.

86. Statistics say that about 1 in 4 blind dates result in a second date. Considering that only about 3 in 100 regular dates lead to an eventual marriage, this puts your chance of success at a depressing 0.75%.

87. This implies that about 1 in 133 blind dates will result in your meeting a future spouse. Good luck with that.

88. Okay now that our biology lesson got derailed, let's quickly review. At this point, we have four haploid nuclei.

89. We have a future **ovule nucleus** bordered by two **synergid cell nuclei** right under the **micropyle** at one end of the cell. Another haploid nucleus is floating in the middle of the cytoplasm.

90. The remaining divisions are solely for the purpose of producing accessory cells to help the egg develop before and after pollination. This division creates two **polar nuclei** and three **antipodal cells**.

91. The two **polar nuclei** fuse and become a **diploid cell** again. When the ovule becomes a mature seed, this will become the starch-storing **endosperm** that nourishes the developing embryo with starch.

92. Corn is a great example of an endosperm in a seed that you see just about every day. A bowl of popcorn is mostly just hot buttered endosperms. The husks are mostly the integument.

93. The **antipodal cells** migrate to the opposite end of the ovule. They have the job of nourishing the developing egg cell with lipids and proteins. They stuff the egg full of nutrient vacuoles.

94. So, in the end, there are three mitotic divisions that produce 8 nuclei, that become 7 cells. The reason only 7 cells form, is that the endosperm fuses the two polar nuclei and forms a single cell.

95. When the pollen grain finds its way to the ovule, it grows a tube cell down the micropyle. One of the sperm in the pollen fertilizes the egg to make an **embryo**, while the other fertilizes the **endosperm** to make a triploid cell.

96. The process of ovule development is shown in the diagram below.

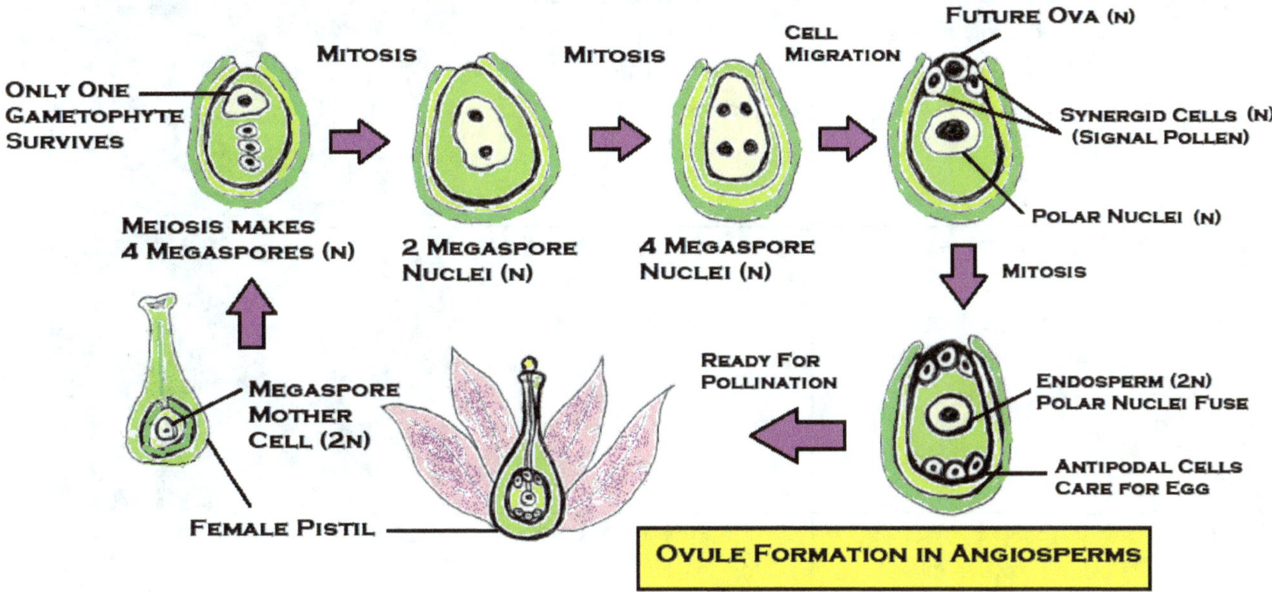

97. How the pollen actually GETS where it is going depends on what strategy the individual plant species evolved. Obviously, plants can't uproot themselves and go out on dates, so something has to do the bidding for the plant.

98. Many plants can **self-pollinate**, while others have genetic limitations on doing so.

99. Many monocots, particularly grasses, are **wind-pollinated**, just like gymnosperms.

100. However, most **angiosperms** rely on **symbiotic relationships** with animal pollinators. After the **ovule** has been fertilized with pollen, the plant moves on with its reproductive plans.

101. The **calyx** and **corolla** are shed, as are the male flower parts, leaving only the **ovary** behind.

102. Inside the **ovary**, the **ovules** develop into **seeds**. The **ovary**, itself, transforms into a **fruit**.

103. The table on the next page summarizes some of the most common symbiotic relationships and gives examples of plants and animals that engage in this type of pollination.

Primary Pollinators	Description	Example	Picture
Honeybees	Bees pollinate flowers that they can see through the UV lenses in their eyes. They also follow scents and chemical trails from other bees. Yellow & purple are favored.	Honeybees pollinate around 70% of commercial fruit crops, including citrus, peaches, apples.	
Butterflies	Many herbaceous plants are specifically pollinated and fed upon by one specific species of butterfly and their caterpillars.	Spicebush is pollinated and fed upon only by spicebush swallowtails. Same for milkweed and monarch butterflies.	
Moths	Some plants are only pollinated by moths nocturnally. Their flowers are white to show up in the moonlight.	Darwin's moth and Darwin's orchid. Sphinx moths and members of the nightshade family.	
Flies	Many plants put out an odor of rotting meat to attract flies as pollinators. Color seems not to be very important.	The Raffanesia plant makes a 3 foot bloom that smells like unwashed butt mixed with dead horses. Flies love it. They save coupons and order it special.	
Birds	Many plants have co-evolved flower structures with the bills of birds that only those birds can pollinate. Unlike insects, they dig the color red.	Hummingbirds love to pollinate the hummingbird trumpet vine....go figure....Many bromeliads are also bird pollinated.	
Mammals	Not the most common mechanism, there are night-blooming flowers (white) that are bat pollinated. Rats, monkeys, lemurs & marsupials also pollinated plants.	Mangos and bananas are often bat pollinated, as are agave plants that they make tequila out of. Sugar gliders & lemurs pollinate plants on accident, drinking nectar.	

B) Diversity of Angiosperm Reproductive Structures and Strategies

1.At first glance, to many people, a flower is a flower, albeit with some variation in shapes, sizes, and colors.

2. However, there is much more to a flower than meets the eye. The **ovaries** and **stamens** of of angiosperms can differ just as dramatically in their structures as the more conspicuous morphology of plants, themselves.

3. The morphologies of male and female flower parts are often very useful in making determinations of how to taxonomically group angiosperms. Let's take a closer look at individual structures, starting with the **ovary.**

4. Flower ovaries are located at the base of the **pistil.** The ovary, itself can be positioned one of three ways as it emerges from the **meristematic** tissues of the **pedicil** connecting it to the **stem.**

5. There are three distinct elevational positions of the ovary in angiosperms.

6. In **hypogynous** species, the pistil is elevated well above the level of the **pedicel** in clear view. Members of the nightshade family, like potatoes, peppers, petunias, and tomatoes have this type of flower. Likewise, legumes (beas, peans, and numerous other allies), brassicas (mustard, cabbages), and some monocots like onions and aloes do.

7. **Perigynous** flowers have a **stigma** that is elevated like the periscope of a submarine, but the ovary, itself, remains tucked slightly into the pedicel, even with its rim. Members of the rose family, such as apples, peaches, cherries, and roses (sorry for the redundancy) have these type of flowers

8. In **Epigynous** flowers, only the tip of the **stigma** emerges from the pedicel, while the ovary remains tucked away inside the pedicel, shielded for protection. Examples of plants that have these type of flowers are honeysuckle, daffodils, cucumbers, pumpkins, and melons.

9. The diagrams below illustrate the concept of flower position as a means of classification.

10. In continuing with the description of differences in the reproductive structures between taxa of angiosperms, we will now look at the number of chambers within individual flower ovaries. These are called **locules**.

11. As you probably know from cutting slices of fruit, not all fruits are structured identically. For now, we will stick with the concept as it relates to the flower, but for more description on their development into fruits, see chapter 5.

12. Plant ovaries may have a single lobe called a **locule** or multiple divided **locules.** In botany, the number of locules in each flower pistil just gets a numeric prefix as a descriptor. This number often varies between taxa.

13. **Unilocular ovaries** have a single chamber that the contains the ovules. The endocarp of the ovary does not divide the seeds up into separate chambers as they develop into fruit.

14. Examples of plant orders that usually have **unilocular ovaries** include legumes and rosids. If you split open a pea pod or slice an apple, there are no fibrous walls that divide the seeds into separate chambers. They are all contained in an undividied single ovary.

15. **Bilocular ovaries** have a single divison in the middle, dividing the flower pistil into two distinct lobes. Examples of plants that have two-lobed ovaries include members of the nightshade family, such as tomatoes and eggplants.

16. **Trilocular ovaries** are divided by the ovary wall into thirds at 120 degree angles. Seeds develop in three distinct chambers, as they do in alliums like onions, garlic, and shallots,

17. **Tetralocular ovaries**, in continuing with this trend, have four chambers, as in brassicas, such as turnips and cabbages. There are also **pentalocular ovaries** as occurs in mallows like okra and hibiscus.

18. The diagram below shows cross-sections of each type of ovarian division.

19. In addition to the position of the ovary on the pedicel and the number of lobes, another difference among taxa is the location inside the **ovary** where the **ovules** attach. There are several different orientations of seed attachment.

20. **Placentation** refers to the location in the fruit where the endocarp attaches to the seeds via the **funiculus**.

21. **Marginal placentation** means that the seeds are attached to a single edge of the ovary. This is commonly seed in peas, beans, honey locust tree pods, and in other legumes. Brassicas and lupines also use this method.

22. **Axillary placentation** occurs when the seeds are directly attached to a center placenta derived from the **ovary wall**. Examples of this arrangement include tomatoes,peppers, and other nightshade fruits, as well as citrus fruits.

22. In **peripheral placentation**, there is a central extension of the ovary wall that divides the fruit in the vertical plane, but instead of being attached at the center, the seeds are anchored to the placenta laterally toward the fruit's **mesocarp**. Melons, cucumbers, squashes, and pumpkins all use this type of seed attachment.

23. **Free central placentation** occurs in the fruits **unilocular flowers**, such as figs, where the ovary wall does not grow through the center of the fruit. The seeds are attached to the lateral walls of the ovary and float in the center.

24. In **basal placentation**, which is common in achene fruits like sunflowers and strawberries, the seed takes up a large volume inside the fruit and is attached to the ovary at its base.

25. The pictures below show these five seed position placentations in angiosperms.

26. An additional difference among different taxa of angiosperms is whether they have simple flowers with a single pistil or compound flowers with multiple pistils.

27. **Monocarpellary flowers** have a single pistil (and ovary). Legumes of all types and mangoes are examples of plants that produce monocarpellary flowers.

28. **Bicarpellary flowers** have two pistils at the center of each flower, and hence two ovaries. It is important to remember that this is independent of the number of locules in each ovary. For instance, nightshades are **bicarpellary** AND **bilocular**. However, cabbages and turnips are **bicarpellary**, but **tetralocular.**

29. In other words, tomato flowers have two ovaries, each with two lobes, while a cabbage flower has a pair of ovaries that have four lobes in each of their pistils.

30. **Tricarpellary flowers** are clusters of three ovaries in each flower. This is a common arrangement of pistils in the flowers of monocots, such as lilies and tulips.

31. **Polycarpellary flowers** have multiple ovaries in the center of each flower. There are numerous different types of polycarpellary flowers, based on the number of pistils in each cluster. Examples include carnations and poppies.

32. The pictures below show examples of species that have each type of pistil arrangement.

MANGO
MONOCARPELLARY

CHILI PEPPER
DICARPELLARY

ORIENTAL LILY
TRICARPELLARY

OPIUM POPPY
POLYCARPELLARY

33. In addition to differing morphologies between ovaries, the **anthers** of different flowers also have a variety of different forms that are used as criteria for classification.

34. The first criteria used for classification is the position of the anthers at the top of the flower **stamen**. There are six different basic designs of flower stamens.

35. **Adnate anthers**, found in primitive dicots like magnolias, attach to the **filament** along the entire dorsal surface.

36. Many **monocots**, such as corn, bamboo, rice, and other grasses, have **versatile anthers**, which have a single point of attachment to the filament on the middle of the dorsal surface. Both ends are free.

37. There are several different types of anther positions in **eudicots**.

38. **Dorsifixed anthers** attach to the filament from the middle of the dorsal surface to the bottom end. Passionfruits and other members of the Malpighiales have these type of anthers.

39. **Basifixed anthers** only attach to the filament at bottom of the stamen. Brassicas, such as kale and cabbages, and mallows, such as okra and cotton, have this type of flower design.

40. There are also two unusual-looking designs found in some angiosperm flowers.

41. **Divergent anthers** are rooted at their bases to the filament, but point away from each other, rather than remaining parallel. Trumpet vines have these type of anthers. **Distractile anthers** , as in pincushion flowers,

point in perpendicular angles to one another, with the horizontal anther smaller in size than the the vertical anther.

42) The six major types of stamen arrangements on the anthers are shown on the next page.

43) Stamens can also be **monothecal**, with a single pair of anthers at the tip (such as in mallows like hibiscus) or they can be **dithecal**, with two pairs of anthers fused to the same filament (such as in onions and radishes).

44) Monothecal and dithecal anthers are shown below.

STAMEN MULTIPLES

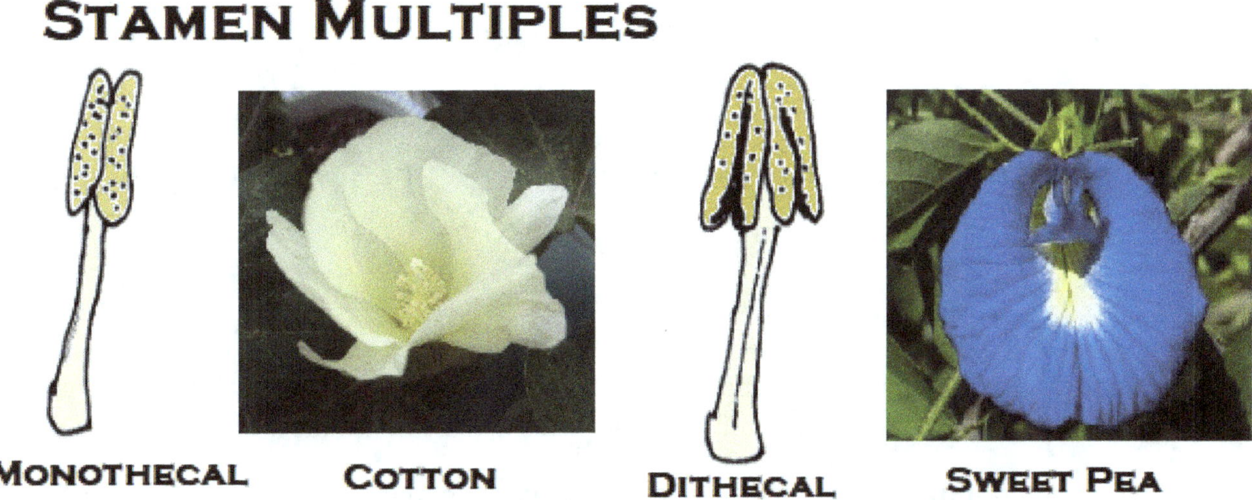

43) The position of the anthers in the flower whorl, relative to one another, also provides a classification character.

44) Didymous stamens have stamens in the cluster, all of equal height. Galium, gardenia, and jasmine and other members of the Order Gentianales have these type of stamen arrangement.

45) Didynamous stamens have a center pair of stamens that elevate above two shorter stamens on their exterior. Many members of **Order Lamiales** , such as olives, mints, and verbenas, usually have these type of anthers.

46) Tetradynamous stamens have two high central pairs of stamens (four in total) and two shorter stamens on the exterior. Brassicas have this type of stamen arrangement.

47) The diagram below shows the three major positional variations among Angiosperm stamens.

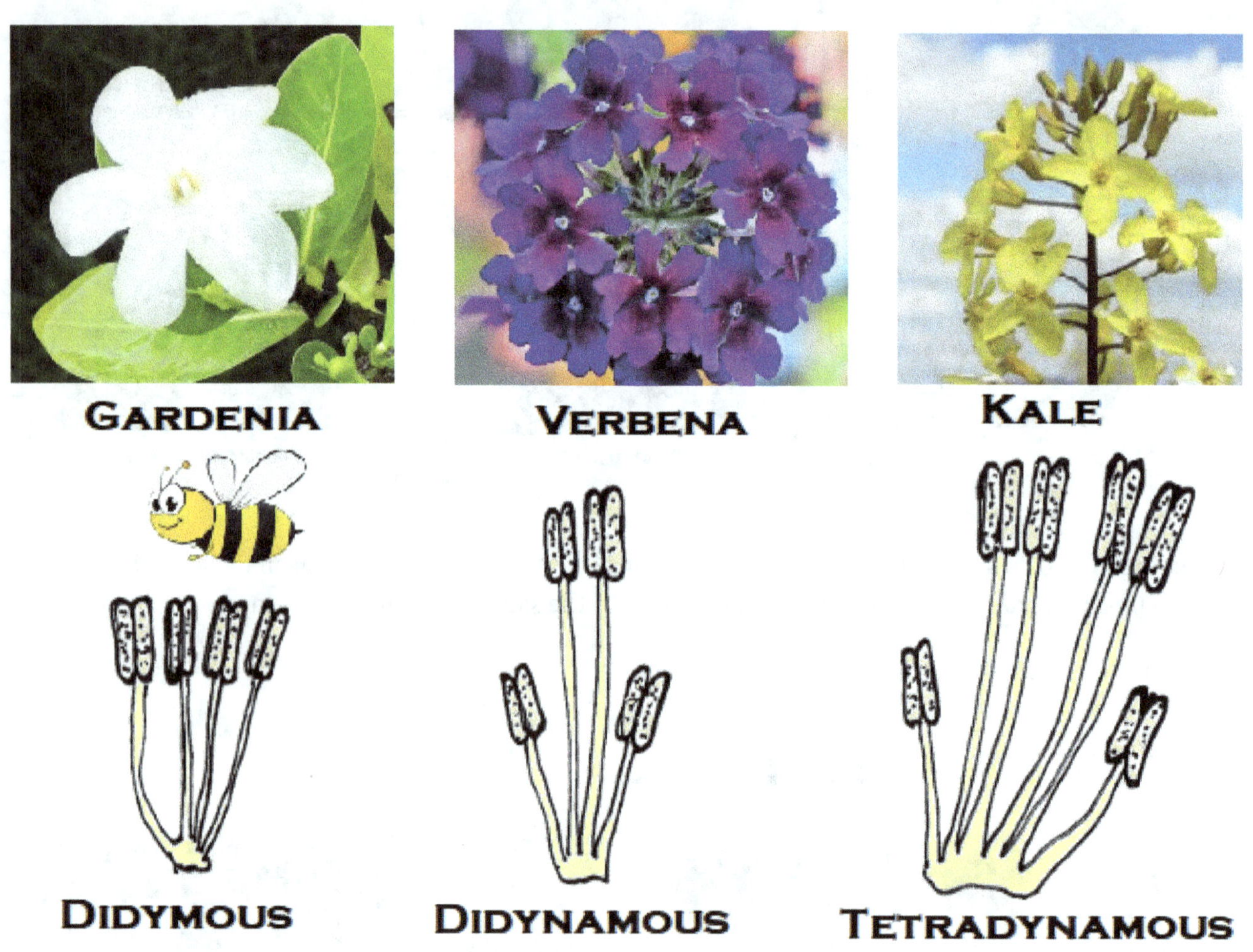

C) The Taxonomic Split: Monocots versus Dicots

1.It is at this point, that we must first acknowledge the large sub-phylum split in **angiosperms** into two different sub-phyla. There are two taxonomic groups of flowering plants that are morphologically very different.

2. We will re-visit **seed development** and **fruit production** after addressing the large taxonomic split between the sub-phylum **monocotyledons** (monocots informally) and the other sub-phylum **dicotyledons** (dicots).

3. **Monocots** typically have fleshy stems, parallel leaf veins, flower parts in multiples of 3, fibrous roots, and scattered vascular bundles if the stem is cross-sectioned.

4. They are most frequently wind-pollinated, though large flowers aren't uncommon. Dry fruits are more common than fleshy fruits, though succulent fruits certainly aren't rare.

5. Examples of monocots include grasses, lilies, yuccas, palm trees, bromeliads, and irises.

6. **Dicots** may or may not have woody stems with ringed vascular bundles, have networked leaf veins, and often have flower parts in multiples of 4 or 5. All known flowering tree species are classified as dicots.

7. The most common reproductive strategy in the dicots is animal pollination. Dicots have a wide variety of different fruit types, many of them fleshy to attract animals to disperse their seeds.

8. With that said, calling all dicots the same thing is an oversimplification.

9. There are actually two groups of dicots. The **basal magnolid dicots**, named for the most primitive flowering plants (magnolias), are the most ancient. They are less evolved than most modern **eudicots.**

10. **Basal magnolid dicots** have single opening in their pollen grains for the sperm to escape, have stamens that resemble the male cones of **gymnosperms**, and often have flower parts in multiples of 3, instead of 4 or 5.

11. **Eudicots** have three different openings in their pollen grains, have stamens that look nothing like cones, and almost always have flowers in multiples of 4 or 5.

12. The **monocots** are believed to have evolved from the **eudicots** between 100 and 150 million years ago, based on DNA and RNA evidence and from suggestive fossil evidence from the Cretaceous period.

13. Apparently, this split happened before some of the changes that made modern eudicots so different than the other two groups, such as the change in floral enumeration and pollen grains.

14. They are so fundamentally different in anatomy and reproductive strategy, that it is important to discuss their differences before moving on to discuss seed formation and fruit development.

15. One reason that these changes may have occurred, is climate adaptations. The earth was much warmer in the Cretaceous period. Back then, **C4 metabolism** would have been much more useful than **C3 metabolism.**

16. **C4 metabolism**, you may remember from the photosynthesis chapter, favors sugar production in hot, dry conditions, since the plant can store a reserve of CO_2 by forming **malic acid** vacuoles.

17. **C4 plants** can keep doing the light reactions of photosynthesis with their **stomata** closed and not lose water during the hottest parts of the day when the sun is most intense. In the tropics, this is quite helpful.

18. It's no coincidence that there are numerous species of monocots that have **C4 metabolism**, because they evolved when the earth was warmer. Likewise some dicots with tropical ancestry also use C4 pathways.

19. Another reason for monocot evolution might have been **niche** utilization. Many monocots, particularly grasses, sedges, and rushes have short life spans and produce hundreds of thousands of seeds as **R-strategists.**

20. Many monocots have short lifespans and can grow in some inhospitable places where dicot shrubs and trees struggle. For instance, the African savanna is full of elephant grass, but doesn't have many trees.

21. **C3 metabolism** seemed to evolve to become much more important as the earth got cooler, particularly in tree species. Playing the long game and using less energy works better in temperate and cold climates.

22. Now let's concentrate on the fundamental anatomical differences between monocots and dicots.

23. As they say, a picture is worth 1000 words. Note the diagrams detailing the differences between the roots, stems, leaves, flowers, seeds, embryos and pollen of monocots and dicots.

24. A descriptive summary chart follows the page of diagrams. Enjoy this plant trivia afterwards.

Characteristic	Monocots	Dicots
Seeds	One cotyledon (kernel) Endosperm plays major role in nourishing developing embryo	Two cotyledons (kernels) Endosperm degenerates quickly. Cotyledons mostly nourish embyo.
Embryo	Leaves and stems emerge from seed as one structure called a coleoptiles.	Stem (hypocotyl) and first leaves (epicotyl) are distinct in appearance.
Flowers	Parts in multiples of 3. Sepals often incorporated into flower with petals.	Parts in multiples of 4 or 5. Sepals not commonly part of flower, but can be in some species.
Fruit	Dry fruits are more common	Fleshy fruits are more common
Leaves	Parallel leaf venation	Networked leaf venation
Herbaceous stems	Scattered bundles of xylem and phloem throughout stalk. Abundant cortex throughout.	Organized vascular bundles in a ring-pattern, with xylem on inside and phloem on the outside. Cortex at center.
Woody stems	No true woody stems. Fibrous tissue mimics wood in some species (palms).	True wood. Xylem deposits in rings at center and crushes out cortex. Phloem develops on outside underneath bark from vascular cambium.
Root system	Often fibrous, numerous, and small.	Often have large taproots or central roots.
Root vascular system	Pith at center with a circle of xylem, circled again by phloem and cortex.	Xylem form an X-shape in center of root. Phloem sits in arms of X. Cortex on outside of root.

25. Take a study break. Take off your orthpedic shoes, pour a large glass of oat milk, grab a couple of prunes, and turn off your hearing aid.   Do a crossword. This is 'you' time, baby.

Eating the Plant Kingdom
Identify either the plant from the food clues.

Answers Page 150

Across

3. Genetically, it is a wild mustard plant with a mutation for a giant terminal bud.
5. Relative of cotton and hibiscus that is deep fried and used in gumbo.
7. Other than anchovies, this bromeliad is the most controversial topping for a pizza.
8. Produced in response to insect damage, the amount of capsaicin in these berries is measured in Scoville units.
9. While its name is a misnomer, this wild grass seed is often mixed with with its true namesake in pilaf dishes.
12. These tart stalks are sweetened and used in pies with berries, but all other parts of the plant are toxic.
14. A pungent, wild native grape found in the Southeastern USA that is prized for jellies, wines, and table fruit.
16. This member of the morning glory family is not a potato at all, but is used as such.
20. Legumes with lens-shaped seeds that are a staple in Middle Eastern food and Indian curries

Down

1. Tequila is made from the sugars in this plant's flower stalk, sisal from its leaves.
2. These fruits are wasp-pollinated. Often dried, they are eaten with brie cheese.
4. A bumpy soccer-ball sized fruit native to the South Pacific. Said to taste like a freshly baked version of its namesake.
6. A pudding made from the root of the cassava aka manioc plant. Often vanilla-flavored.
10. The honeydew princess and watermelon prince had to get married in church, because their parents insisted that they....
11. Technically not vegetables, these gigantic berries can grow to over a ton in weight.
13. Common in the Appalachians, a wild relative of garlic often picked for gourmet dishes.
15. The petals of this flower are edible and its oils are used in Indian desserts.
17. A starchy giant banana used for snack chips and baked in savory dishes like a potato.
18. By consensus, the most vile fruit known. Smells like unwashed gym clothes and beloved by orangutans.
19. A member of the mint family, its many cultivars include sweet, Italian, Thai, purple, and lime.

Chapter Four

Diversity of Angiosperms: The Monocots and Dicots

A) **Monocot Diversity and Taxonomy**

1.Monocots have been extremely successful at colonizing many different habitats on earth, from temperate and tropical grasslands, to rainforests, and even polar ice caps and underwater.

2. Let's quickly recap the typical characteristics of **monocot** plants. While it is thought that they evolved from a basal dicot, probably something similar to the water weeds found in **Order Ceratophyllidae**, they are a distinctly different evolutionary offshoot of the flowering plants.

3. Monocots are fundamentally different reproductively, in that their pollen is **monosulcate** (one groove for pollen tube exit), their flowers have parts in multiples of 3, and most supply their embryo with a single **cotyledon** filled with nutrient reserves.

4. The leaves of monocots also have distinct characteristics. Most monocots have parallel leaf veins and commonly have **sheaths** at the base near the petiole. C4 metabolism is more common among their members, which is reflected microscopically in the presence of a single type of **mesophyll** and with enlarged **bundle sheath cells**.

5. The stems of monocots lack the ability to produce true wood fibers and show scattered **vascular bundle** distribution in microscopic cross-sections. The majority are herbaceous, but even the ones that show secondary growth use hardened **ground tissues** like **collenchyma** and **sclerenchyma**, rather than layers of concentric **xylem** (vascular tissue).

6. Monocot roots show concentric vascular bundles with a layer of **pith** in the center. Typically, they are fibrous.

7. The chart below resummarizes the main features of monocot plants.

8. Phylogenetically speaking, the most primitive monocots are the **Order Acorales**, which are commonly called the 'sweet flags'. They lack an **endosperm** in their seeds and have primitive flower structures. The arums of **Order Alismatales** evolved **tepals** (petals and sepals that are relatively indistinguishable), but still lack the endosperm.

9. The more advanced monocots have an endosperm in their seed, but vary in their flower structures and at the cellular and biochemical level. Generally speaking, the advanced monocots are divided into the **lily-like monocots** and into the **commelinid monocots** because of these difference.

ROCK SCHOOL QUIZ ANSWERS FROM PAGE

1) Van Halen's contracts required a bowl of M&M's with every brown candy removed. If this wasn't done, the promoter forfeited payment for the show. This was a safeguard for the band to make sure their detailed contract had been fully read, since the set-up to achieve their sound was complicated.

2) Rod Morgenstern was (and is) the drummer for Winger. In spite of idiotic lyrics to some of their songs, he had few skill rivals in the hair metal genre and became a percussion professor. While he didn't become a professor, guitarist Reb Beach hold equal regard for musicianship.

3) Axl Rose was irate that a fan was taking pictures of him. When security was slow to act, he dove off the stage and tried to confiscate the camera himself. A riot broke out when he punched the fan.

4) David Coverdale of Whitesnake decided he liked what he saw of Tawny Kitaen from videos like 'Is This Love' and married her, so she could slide around on his expensive luxury cars full time.

5) David Lee Roth entered the convenience store in full headhunter regalia carrying a spear. He looked both ways suspiciously before demanding a 'bottle of anything......and a glazed donut.....TO GO!'

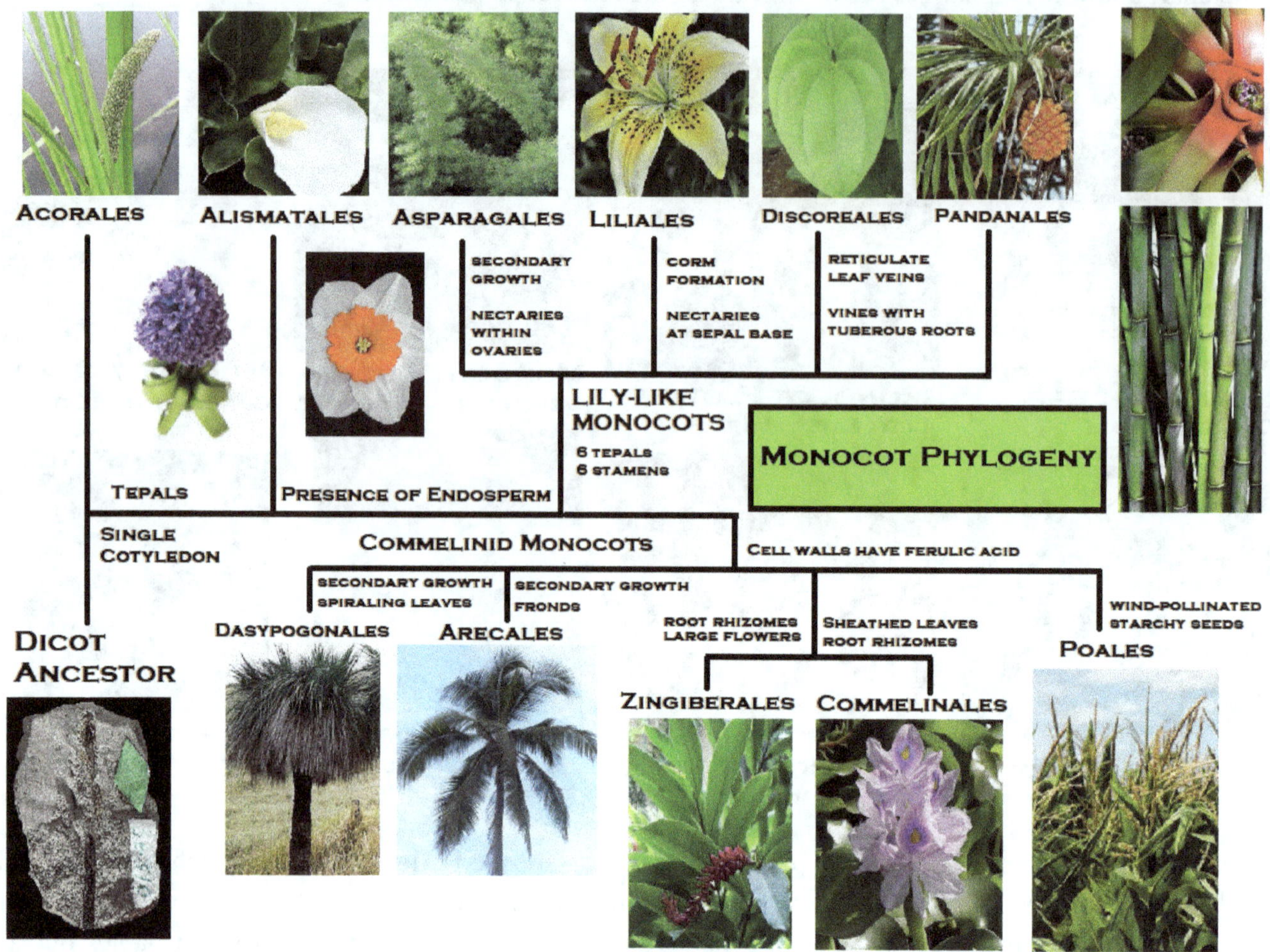

10. The diagram above is a representative cladogram of the phylogeny of monocots. Keep in mind that there are alternative models and that biochemical evidence continues to accumulate that might possibly alter existing models.

11. Now let's turn our attention to a more detailed look at the major orders of monocot plants.

12. At present, there are around 60,000 known species of monocots in 11 different orders. The most numerous, in terms of species, are the orchids (more than 20,000 species) and the grasses (more than 10,000 species).

13. We will briefly cover some of the monocot orders and describe some of the members of each.

14. One of the oldest group of monocots are the sweet flags. They are found in the wetlands of Asia. Due to primitive pollen and flower structure, they can't really be classified with anything else.

15. The sweet flags, members of the **Order Acorales**, are shown below. They have spadix-shaped flower clusters with the stamens below the level of the petals, and sword-shaped leaves that lack brackets growing into the stem.

16. They lack well-developed petals or sepals on their flowers and their seeds have no **endosperm** tissues.

SWEET FLAG   **GRASSLEAF FLAG**

ORDER ACORALES (SWEET FLAGS)

DIVERGED FROM OTHER MONOCOTS LONG AGO. ONLY 39 RECOGNIZED SPECIES.

MULTIPLE TINY FLOWERS GROW FROM A FLESHY SPIKE CALLED A SPADIX.

BLADE-LIKE LEAVES THAT PRODUCE TOXIN CALLED ASORENE, AS WELL AS OILS.

SOME SPECIES ARE TETRAPLOID OR HEXAPLOID.

17. The arums of **Order Alismatales**, falsely called lilies, by some, include the misnamed arum lilies, calla lilies, peace lilies of florist gift basket fame, a number of freshwater swamp weeds, and the only TRUE marine plants in the world, such as turtle grass (not really a grass).

18. Manatees largely manage to get themselves run over by boats because they are looking for plants in the Order Alismatales. In their own way, manatees fight dangerous consequences of addictions to plants, just like some of us. While the means of calamity may differ (driving under the influence versus being driven OVER), the outcomes are no less tragic.

19. Let the graphic novel representation serve as a stark reminder to any manatees reading this text. Even better, if there are younger sirenians (boyatees) reading, it is better to learn to prevent dangers at a young age, rather than letting life teach you.

IT WAS JUST ANOTHER PEACEFUL DAY IN SUNNY FLORIDA....UNTIL......

20. Unlike most other monocots, their seeds lack the endosperm kernel and their embryos emerge underwater, already having developed chlorophyll. This group is dependent upon being submerged in water to develop.

21. Their flowers grow from a single stalk yet they are **dioecious**, because there is a set of **staminate** flowers at the tip, while **pistil** bearing flowers are located further down on the flower. They have a hood like bract called a **spathe**. This is the white petal-like thingamajig on a peace lily. Some members of the order have **bilocular** ovaries.

22. The seeds of most members of this order are **achenes** with **basal placentation**, though many superficially resemble berries and have red skins. Many, like calla lily 'berries' are loaded with oxalic acid and will send you to the E.R. if you eat them.

23. The giant arum, a member of this group, produces the largest single flower in the world, with the pistil reaching 9 feet tall. The 3 foot wide bloom smells like a sun-bloated dead cow, attracting flies to pollinate it.

24. Several representative members of **Order Alismotales**, are shown in the next chart.

25. Now let's make our way through the **lily-like monocots**, one of the two more advanced groups that have an endosperm in their embryos. Additionally, the lily-like monocots typically exhibit 6 **tepals** on the corolla of their flowers and have 6 **stamens**, though there is variability in ovary structure.

26. The lily-like monocot clades includes **Order Asparagales** (asparagus and allies), **Order Liliales** (true lilies), **Order Discorales** (true yams and allies), and **Order Pandanales** (pandanus screw-pine).

27. The largest order of monocots is the **Order Asparagales**, which contains more than 36,000 species. Many familiar plants, such as orchids, garlic, asparagus, onions, aloe, irises, gladiolas, and day lilies belong to this group.

28. Members of this order generally have flat leaf blades and their stems are also flat with the leaf. Many have leaves that grow outward from their sheaths in a spiraling whorl.

29. The majority of the species in this order have dry encapsulated seeds and many of them grow leaves from a center whorl.

30. The flower pistils of most members of the order are **hypogynous**, rising above the level of the petals. Ovaries are **trilocular** and the flowers are **tricarpellary.**

31. Like the more distantly related palms, some members of this group can grow false woody stems by adding layers of vascular bundles outside the center meristem. Again, rather than true wood (vascular tissue), these fibers are formed by ground tissue.

32. **Orchids**, in terms of species number, make up 28,000 of the 36,000 species in this group. Found throughout the tropics, many grow as **epiphytes** on other trees. They are very important to the ecology of tropical forests.

33. Other than a few species of vanilla, few are commercially important as anything other than difficult-to-keep ornamental houseplants that you to your mom for Mother's Day, dooming them to death within weeks.

34. It's not really her fault though...most orchids are extremely dependent on very sensitive root **mycorrhizae**. If the soil, humidity, nutrients, and temperature aren't perfect, both kick the bucket in succession.

35. The vegetable aisle of the grocery store is full of members of this order. In addition to the namesake asparagus, all members of the allium clan belong to this order. Onions, garlic, leeks, and shallots are all members of Asparagales.

36. Since they grow in climates that often have long winters, they evolved true **bulbs.** Bulbs are a cluster of fleshy starch-filled underground leaves that grow out of the same meristem as the photosynthetic leaves. The roots and main shoot can wait out the winter and survive on food reserves.

37. Additionally, alliums manufacture stinky sulfurous diallyl compounds to discourage burrowing animals and herbivores from digging them up and eating them. They make you cry if you chop a real stinker, but otherwise, we ignore the warning and add them to our spaghetti, tacos, shrimp scampi, and pizza.

38. Many of the Asparagales can be purchased in the landscape aisle of your mega hardware vendors for outrageous prices if you don't know how to trade plants online or grow them yourself at much cheaper prices.

39. For instance, bearded irisis and daffodils, flowers synonymous with Spring, are experts at asexual reproduction. Daffodils, like the alliums, wait out the winter as underground bulbs, while irises grow **corms** (stem nodes full of starch) instead. Both initiate budding into extra bulbs and corms when the extant ones grow too large.

40. Furthermore, corms can be **fragmented**, allowing multiple new plants to be started.

41. Two more familiar plants in this group are sought after for the fluids in their sap. Aloe vera contains glycoproteins with anti-inflammatory properies, while agave flower stalks are full of sugars that can be fermented to make a bottle of liquid regret...aka DAH-Nuh-Nuh-Nah-Nuh-Nuh-Nah-Nah! Tequila!!

ALOE VERA   **CATTLE ORCHID**   **RED ONION**   **VANILLA ORCHID**

BEARDED IRIS   **DAFFODIL**   **ASPARAGUS**   **GIANT AGAVE**

42. Representative members of **Order Asparagales** are pictured above.

43. The true lilies, members of the **Order Liliales**, are among the most instantly recognizable monocots, with their prototype parallel leaf veins, prototype monocot flowers, and herbaceous stems.

44. Like most monocots, the Liliales have **trilocular ovaries** and **tricarpellary** flowers with **hypogynous** pistils.

45. The **petals** of most Liliales are often indistinguishable from the **sepals** and they often have nectar compartments at their bases directly below the stamens. This design is deliberate, as it forces bees and other pollinators to dust themselves with pollen to feed, thereby spreading gametes to the next flower.

46. Most members in the lily order store large quantities of starch in underground **corms** or **bulbs**, allowing them to wait out winters or dry spells in dormancy as perrenials that return each growing season.

47. As you might expect, having starchy underground food reserves can attract animals that want a meal. For this reason, many members of this order make toxic compounds to protect themselves.

48. **Colchicine**, which interferes with spindle fiber disassembly in **mitosis** and **meiosis** is a toxin that can freeze cell division. If an animal eats enough of it, the toxin can be lethal. It is produced by crocuses. Animals that consume crocus bulbs will suffer profound vomiting and diarrhea at minimum. Organ failure failure and death can follow.

49. Lily-of-the-valley produces toxic **glycosides** that prevent the sodium-potassium pump from getting rid of sodium in cells. This draws excess water into cells and creates osmotic shock. This has toxic effects on many body systems, but is particularly threatening to the heart and kidneys.

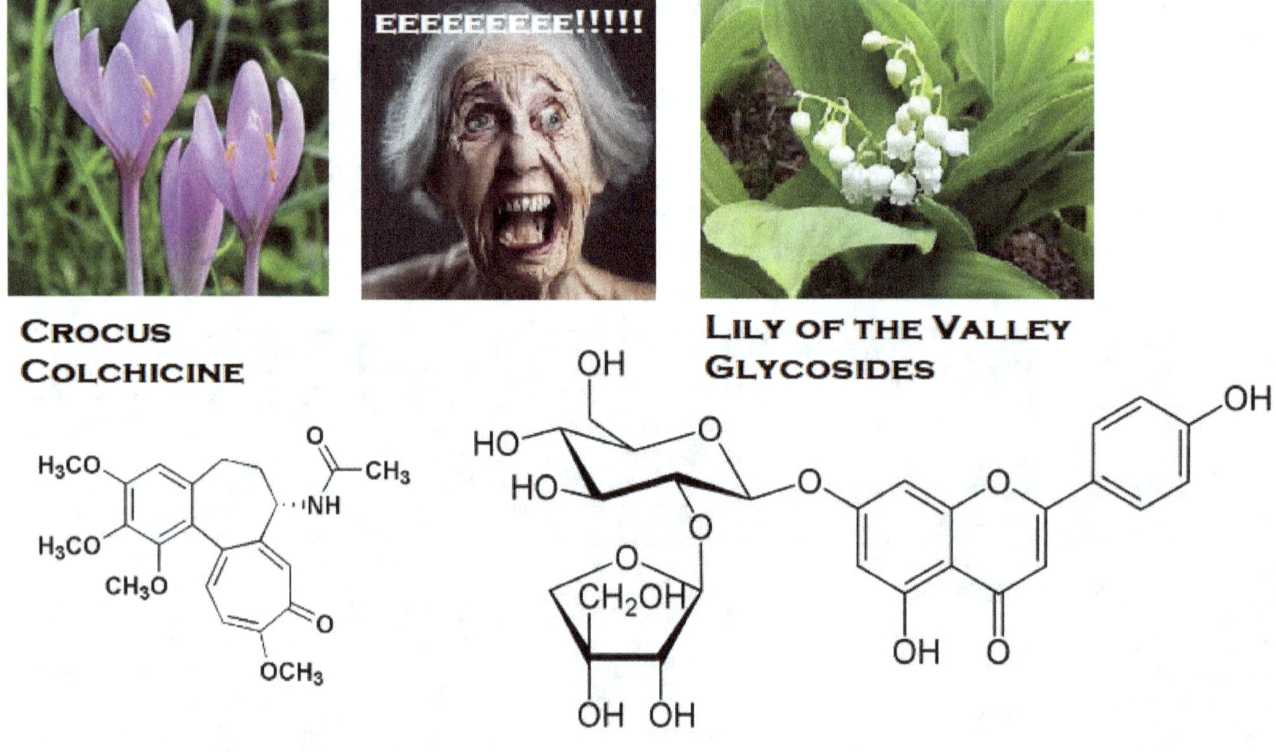

FAMOUS PEOPLE WITH PLANT NAMES....Name the person described. Answers p. 64
1) Master pastry chef and elegant British senior lady who hosts 'British Baking Show'
2) Sold 100 million plus albums as the front man with Page, Jones, and Bonham.
3) Last name shared by NBA point guard Derek, 2nd Baseman Pete, and rocker Axl.
4) First name of Scrooge's niece, Uncle Jesse's niece, and Star Wars star Ridley.
5) Former tough guy U.S. president and military officer known as 'Old Hickory'.
6) Arboreal actor who was in numerous Star Trek, Jack Ryan, and Wonder Woman films.
7) Flowery real first name of Tommy Lee's ex and the fictitious Miss Kensington.

50. Diagrams of the molecular structure and source of each toxin are shown above, along with what can happen to you if you eat these flowers.

51. Almost everyone has had some sort of lily planted in their yard or been the purchaser or recipient of a flower bouquet, containing said lilies. This begs the question of how this expensive and bizarre tradition even started.

52. We all know that women like flowers, but how did this social custom even start? While ancient Romans, Greeks, Egyptians, and Turks may have all helped this ritual get started, we mostly have the Victorians to blame.

53. Victorians thought they were clever, but in reality, they were basically socially inept. Thanks to this issue, millions of high schoolers are tortured every year, as they are forced to slog through such 'classics' as Jane Eyre.

54. It takes hundreds of pages of boring insinuations and female friends gossiping in guarded innuendos about some Duke, Earl, trashman, whatever, who is courting one of them in novels like Jane Eyre or Emma or whatever.

55. In the meantime, almost nothing of substance actually happens. No explosions, no animal attacks, no zombies.

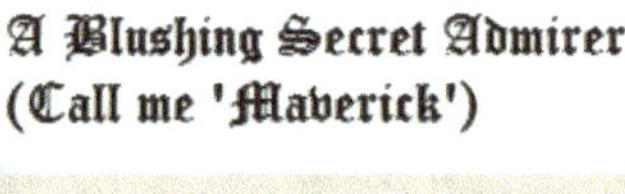

56. Instead of taking the bull by the horns like a man, the nobleman in question fiddles around for 500 pages, before finally telling the Duchess, Countess, Princess, cafeteria lady, whoever, that he has the hots for her.

57. Such Victorian men sent flowers to symbolize various emotions, rather than actually talking to the woman. In a way, it's not much different than sending emojis via text. It really sort of borders on being stalkerish.....

58. Lilies and their relatives have been frequent victims of this tradition that continues today. They are easy targets, given the large showy flowers of most of the nearly 1800 species known to science. Lilies are common across the Northern hemisphere from the sub-Arctic to the tropics.

59. The chart below shows some of the many members of this diverse order of plants.

60. The next lily-like order of monocots is less familiar to most people, but they are found circumglobablly.

61. **Order Discoreales**includes the true yams, a number of other vining relatives, and even a few parasitic heterotrophic members that discarded their chloroplasts in favor of stealing nutrients from the roots of hosts.

62. Many members of this order tend to have deep **tuberous** roots capable of sustaining the plant through long periods of drought, such as the many species of African yams. This has made them desirable for cultivation for food in areas where agriculture can be difficult.

63. The leaves ofmembers of this taxa are unusual among monocots,since they have **reticulated leaves**, rather than exclusively parallel leaf veins. Whether this was re-evolved as a product of convergent evolution or retained from ancestral dicot genes,this trait isuseful, since many of its members have broad leaves for trapping sunlight under deep forest canopies.

PLANT PEOPLE ANSWERS

1) Mary Berry	2) Robert Plant	3) Rose	4) Daisy (Duck and Duke)
5) Andrew Jackson	6) Chris Pine		7) Heather (Locklear and Graham)

64. The diagram below shows several members of the Order Discoreales.

ELEPHANT'S FOOT YAM   CAUCASUS YAM ROOT   BLACK BRYONY

65. Members of the **Order Pandanales** include pandanus screwpines, Panama hat plants, cattails, and a number of other fibrous woody tropical shrubs and vines. Most have thick blade-like leaves.

66. However, there are some exceptions to the norm, such as the South African desert plant species in genus Xerophyta, which have reduced leaves and thickened stems as adaptations to their arid habitat.

67. The strangest members of the taxa are the heterotrophic members of the family Triuridaceae, which lack chlorophyll altogether.They rely solely on **mychorrhizal fungi** to feed them. It is assumed that the fungus gets an increased amount of nutrients and moisture by using the plant roots on the other side of the symbiotic relationship.

68. Proving that morphology isn't always a reliable predictor of phylogeny, it is only by flower structures and DNA analysis that it is apparent that plants like Lacandonia belong to Order Pandanales

69. As mentioned, what gets these plants classified together is the anatomy of their flowers. Most members of this order have four **didymous** stamens, two **bicarpellary trilocular** pistils, and distinctly tiny embryos with very large **endosperms** (relative to their size) that can keep the embryo alive for a long time.

70. Some member of the order are wind-pollinated, such as the screw pines, while others like Freycinetia have large showy flowers to attract birds and bats that pollinate them. The vellozias are pollinated by bees and butterflies, while many members of the vining Stemona family are fly-pollinated and their flowers put o n a stinkfest to attract t

71. Pandanus screwpine has such large (and buoyant) seeds that they can float for months on the open ocean before finding their way to a desolate shoreline on some South Pacific island hundreds of miles from where they started their journey. The large endosperm is needed for this journey

72. Oddly, the arum lilies (not true lilies) also belong to this order of plants. These are good illustrations of the fact that morphology of the sporophyte isn't always a reliable way to tell relationships.

73. The diagram below shows representatives of the Order Pandanales.

74. One of the best known and most beloved and economically important groups of monocots are the palms. They belong to **Order Arecales**. There are more than 2600 species of palms, many commercially important. Coconuts, palm oil, dates, and rattan are all major markets, as are dozens of landscaping species. See the diagram below.

A) COCONUT PALM (NUTS, LANDSCAPING)   E) ROYAL PALM (LANDSCAPING)
B) DATE PALM (FRUIT, LANDSCAPING)   F) RATTAN PALM (WICKER, FIBER)
C) OIL PALM (FOOD ADDITIVES, FUEL)   G) ACAI PALM (BERRIES)
D) CABBAGE PALM (HEARTS OF PALM)

75. Palms are ubiquitous around the world within the tropics and may grow far into milder temperate regions.

76. Palms, like all other monocots, cannot produce true wood fiber, instead initiating fibrous growth from the many vascular bundles scattered throughout their stems. These are mostly ground tissues, such as sclerenchyma and collenchyma.

77. Some palm species grow a crown of leaves from one terminal **meristem** region, while others grow clumps of clones from **lateral meristems** at the base of the roots. Only a few species like the doum palm can branch.

78. Palm flowers, like those of the arums, usually grow in a **spadix** and often have separate regions of male and female flowers on the spadix. **Bracts** protect developing flowers. Typically, they have **trilocular ovaries** and **bithecal anthers**, but there are variations among families.

79. Most palms have **tricarpellary** flowers, but there are exceptions among old world families.

80. Palm fruits are **drupes**, even coconuts and acai 'berries'. They have a single seed in each ovary, which can either be dry and indehiscent (such as a coconut or oil palm 'nut') or fleshy, as it is in palmettos or acai palms. Oddly, their seeds lack starch and they use oils and alternative carbohydrates instead.

81. Palms with fleshy fruits use the drupes as bribes for animals (dates and acai berries) or long nutrient stores for seeds that may lay dormant for many years as they float across oceans and seek out a place to sprout (coconuts).

82. Palm leaves are either **palmate** fan shapes or **pinnate** fronds with many **leaflets** emerging from a central stem. They grow from a sheath originating in the tissues of the stem, growing outward from the center of the crown from tissues known colloquially as the 'heart', as in hearts-of-palm salad.

83. The diagram below shows some of the distinctive traits of palms.

84. The **Order Dasypogonales** includes most Australian members. The existence of this order is not recognized by many taxonomists who split the various species in **Arecales** and **Asaparagales** instead.

85. For now, we will pretend like I'm better than those 'expert' taxonomists with DNA sequencers, banks of computers, phDs, and worldwide noteriety. After all, I AM a high school teacher AND a college adjunct. Do

those guys drive a Bentley to work?Do they have a solid mahogany desk? Do they vacation in Fiji? Do they have adoring fans and shoo away the papparrazi?

86. Like the marsupial mammals, most members of **Dasypogonales** evolved in isolation in the land down under (can't you hear the thunder?! you better run, you better take cover!). Members of the group include the odd grass trees such as Kingia and Xanthorrhoea, tinsel lilies (a misnomer), and the namesake flowering Dasypogons

87. The flowers of this group are **dioecious** and may have branched **inflorescence** with multiple flower heads, (such as the tinsel lilies) or they may grow in a tight cob-like structure with multiple flowers, such as in gras s tree

88. There are two whorls of 3 **tepals** in the corolla. Flowers have two whorls of **tetradynamous stamens** and **trilocular ovaries** that mature into dry **indehiscent capsule** fruits.The seeds have a very large **endosperm**, as you might expect in an arid habitat.

89. The leaves of most members(especially grass trees) are long and skinny, with tough grass-like blades and parallel leaf veins. Spiraling is common. In grass trees, clusters of dead fibrous leaves form a protective trunk-like facade over the stem.

90. This is distinctly different than the way palms support themselves (fibrous ground tissues inside the stem), which is one reason I have chosen to go with this system of taxonomy

91. Dasypogon stems themselves are often thickened and trunk-like as an adaptation to withstanding the arid Outback ,but may be thinner in the herbaceous members (tinsel lilies). Likewise,the same trend tends to be true for branching (absent in the larger members and present in herbaceous members). There is no true **secondary growth**.

TINSEL LILY   **XANTHORRHOEA**   **KINGIA GRASS TREE**   **DASYPOGON**

91. In terms of anthropological and economic importance, every other group of monocots pales in comparison to **Order Poales**, the grasses. In addition to serving as feed for livestock worldwide, grass crops include rice, corn, barley, oats, and wheat. Somewhat unique among members of this group is the incorporation of large amounts of silica into their cell walls.

92. With nearly 22,000 known species, Order Poales also includes close cousins, such as bromeliads, sedges, rushes, and many others that are NOT true grasses, but share a lot of similar derived characteristics.

93. Members of this order have small, simple flowers that are usually covered with a modified protective leaves called **bracts.** It is common for many flowers to grow off of a main stem on lateral side branches. This arrangement is called an **inflorescence.**

94. The anthers are **versatile**, with a thin attachment to the filament in the middle, allowing them to stick up and broadcasat their pollen in the wind.

95. Unlike most other flowering plants, most members of Order Poales rely on wind to pollinate them (though this is not true for most bromeliads). Their seeds typically have large starch-filled **endosperms** that allow the embryo to wait out harsh conditions.

96. The diagram that follows only shows a small cross-section of the many thousands of species that belong to this order. Almost no one makes it through a day without somehow being impacted by these monocots.

97. Bamboo is used for hardwood floors, feeding pandas, and Chinese stir fry. Pineapple is used for fruit salads and questionable pizzas. Name a grain, and there's a great chance it is a grass or close relative and that it is used for food by people or livestock.

98. Corn leads the way in world crop production, with 1.16 billion metric tons per year, followed by wheat (765 million), rice (496 million), and barley (157 million). Sorghum, oats, and rye also account for huge volumes.

99. The diagram below shows just a few of the many thousands of species in this large order.

100. The **Order Zingiberales** contains a number of very distinctive tropical-looking plants, such as bananas, ginger, heliconias, bird-of-paradise, turmeric, cardamom, and galangal. Large, wide, blade-like leaves are common throughout the order.

101. Typically, the midrib of these leaves is rounded and fleshy and emerge from a fleshy stem in whorls.

102. Likewise, the petals of flowers emerge in whorls as well. Often, these are large cone-shaped compound **infloresences** with many pistils fused together in the middle. Often, they are **hypogynous** and full of nectar. The individual clusters of pistils are **tricarpellary**.   Pistils are **tetradynamous**. Often, there are spiraling **bracts** that surround the flowers.

103. Many of the Zingiberales produce showy flowers and aromatic compounds that give them distinct uses as spices. There are many types of ginger and galangal used across the Orient and India for flavorings. Bananas and plantains, of course, form a large portion of the diet throughout the Tropics.

104. Since most of these plants are adapted to the understory, wide leaf blades are the norm. As a result, many members of the taxa (such as bird-of-paradise and traveler's tree) are used in landscaping to create a tropical look.

105. Another common trait of members of the Zingerberales is the presence of **underground rhizomes** for food storage. These rhizomes are capable of **budding** many times to produce asexual clones of their mother plants.

106. The picture above shows several features common to most members of the taxa.

107. The next diagram below shows a few representative members of this large order of around 2,600 species.

108. Members of the **Order Commelinales** are mostly tropical herbaceous perennials with short-lived showy flowers. One distinctive trait of this group is that they have large amounts of ferulic acid, a phenolic compound, in their cell walls.

109. Ferulic acid is similar to the compounds in sun screen providing the plant with UV protection, since many members in this group are adapted to live in direct, intense tropical sun.

109. **Rhizome**-like roots are common in both terrestial species, such as in spiderworts, or in the aquatic water hyacinth species that clone themselves so effectively from the rhizomes that they invasively choke off water ways throughout the tropics.

110. Depending upon the family of plant in consideration, members of the Commelinales have a wide variety of flower structures. While they are more commonly **dioecious**, there are **monoecious** species within the group. There is also variability in the structures of the flowers themselves. However, rounded petals and blue or purple **anthocyanin** pigments are common.

111. Unlike some other members of the monocot clade, the seeds of members of this taxa are loaded with a large starchy endosperm to ensure long-term survival of the embryo.

112. Genetic evidence seems to point to members of the Order Zingiberales (ginger) as their closest relatives, but their common cell wall structure lends their namesake to all of the **Commelinid monocots.**

113. Several members of the Commelinales are shown below.

B: Dicot Diversity and Taxonomy

1.Dicots, according to modern genetic evidence and fossil records, are an older group of angiosperms than the monocots. Overall, there are around 175,000 species of dicots, outnumbering monocots nearly 3:1.

2. Dicot classification has been, at best, a challenge. It is now mostly accepted that there are two fundamental groups of dicots. These are the more primitive **basal dicots** and the more recently evolved **eudicots**.

3. Some of the **eudicots**, however, are also very ancient and retain primitive features that make them look like plants that haven't fully graduated from gymnosperm status.

4. Since this classification scheme can get confusing fast, the diagram below will serve as a reference point for the profiles of plant orders that follow.

5. Keep in mind that DNA analysis, which you can't see, is a large contributing factor to this classification scheme.

6. While there ARE morphological differences between the primitive groups and the eudicots, most physical characteristic are not very useful as obvious pieces of evidence for an evolutionary split.

7. Molecular differences in DNA, chloroplast DNA, and rRNA provide a lot of the basis for their clssification. Plants are particularly adept at **convergent evolution**.

8. The diagram below provides a schematic for dicot evolution and classification.

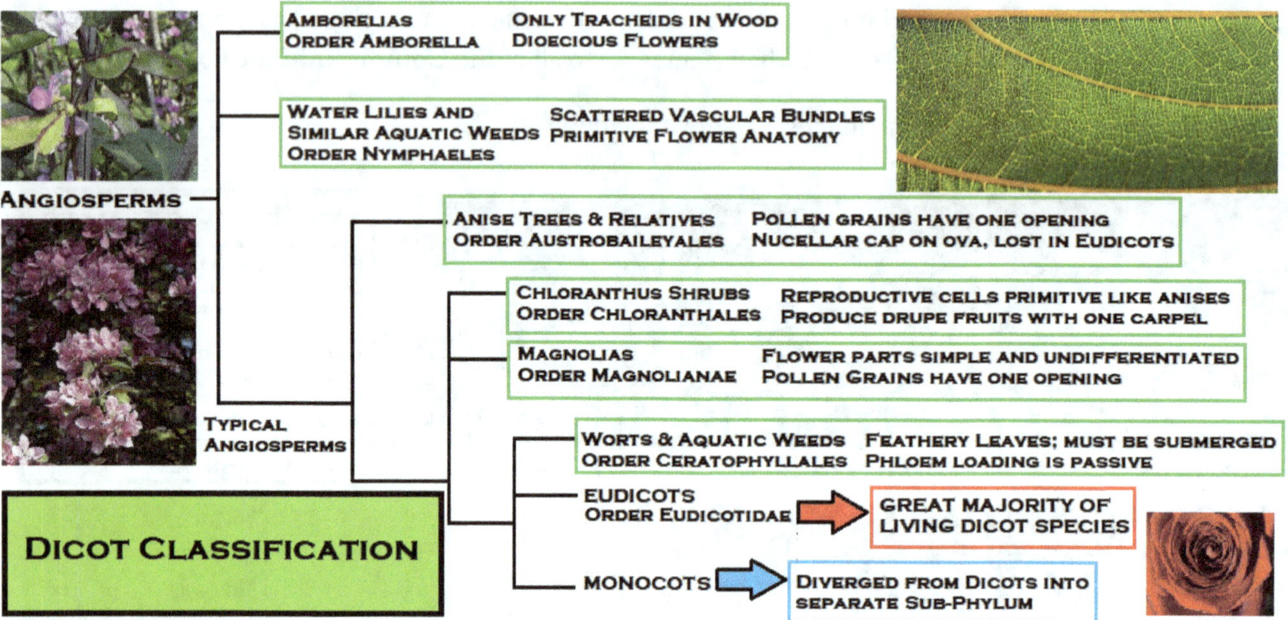

9. The earliest angiosperms probably had many characteristics that were commonly seen in the more primitive gymnosperms, as cones evolved into flowers, wood fibers became more efficient, and fruits developed.

10. Members of the **Orders Amborella, Nymphaeles,** and **Austrobaileyales** are the **basal dicots**. They retain characteristics not typical of modern dicots, or evolved in a different direction.

11. The amborellas have only **tracheids** in their wood and dioecious flowers. Water lilies have scattered vascular bundles, evolved independently of monocots. Anises have very primitive gametophytes that lack many advances of other flowering plants.

12. The flowers of many basal dicots, such as amborellas, have whorled spiraling **tepals**. Unlike higher dicots, there is no obvious physical difference between petals and sepals, so they are collectively called tepals.

13. While flower structure varies among the basal groups, they often grow on spiraling infloresences. While the plants themselves are **dioecious**, they often have separate male and female flowers with a large number of **carpals** or **stamens.**

14. While members of **Orders Chloranthus, Magnolianae,** and **Ceratophyllales** are more evolved than the aforementioned groups, their flower anatomy and pollen structures are very basic in comparison to more advanced dicots.

15. For instance the beloved magnolias, have primitive spiral flower structures surrounded by white tepals. There are numerous **carpels** in the center of the flower, surrounded by a spiral of numerous **stamens** with **adnate** anthers.

16. This type of structure is less common in more advanced dicots, which tend to have fewer carpels in single flowers and distinguishable petals and sepals.

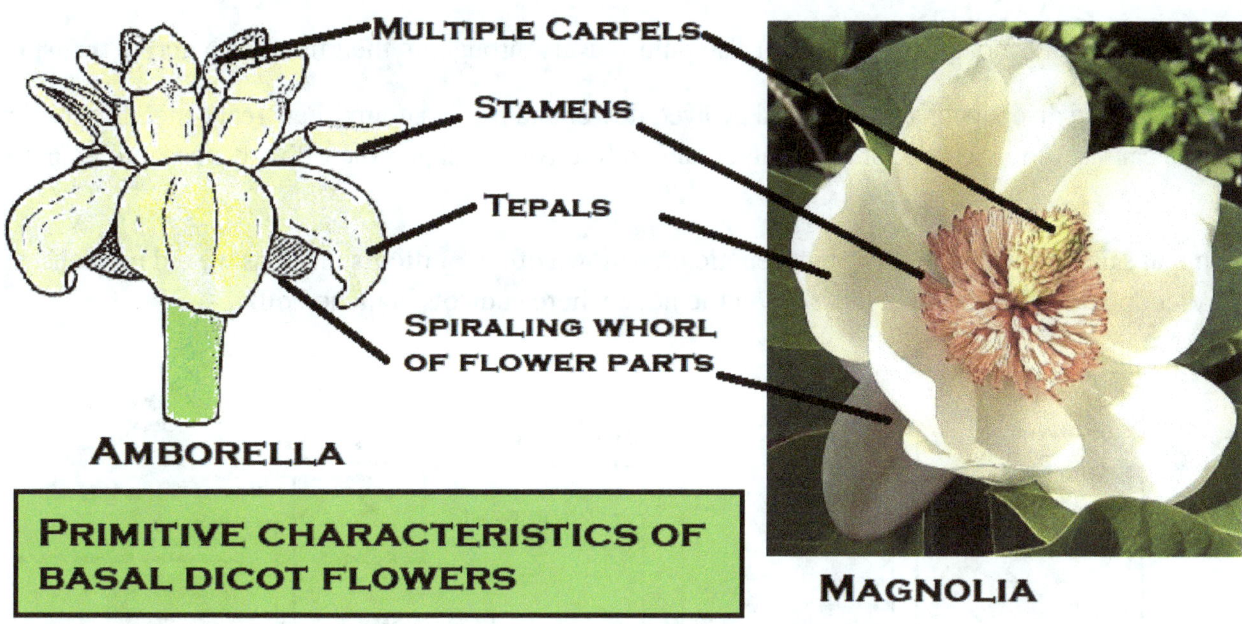

17. The diagram below above amborella and magnolia flowers and points out some of their primitive features.

18. Most of the **basal dicots** are evergreens, except certain magnolias and tulip poplars. **Deciduousness** mostly re-evolved much later in the **eudicots**. **Eudicots** can also pump sugars actively into their phloem, while many basal dicots rely on passive transport and osmosis to move substances.

19. In the interest of keeping things brief, a diagram with some representatives of each of these primitive basal orders is shown below. Many species are also likely to be relatively obscure to readers, having only Latin names.

20. Now we move on to the largest extant group of plants, the **eudicots**. The eudicots have **tricolpate pollen grains**, meaning that there are three openings in the grain and parts in triplicate.

21. Wood fibers of eudicots contain high percentages of **vessel elements**, increasing their conducting efficiency. Rather than being tapered wedges like the tracheids of conifers, they bridge one another to form pipe-like structures.

22. The **phloem** of most of these species can also pump sugars throughout their tissues via **active transport**.

23. There is a huge diversity of advanced and evolved flower and fruit anatomy. Few real generalizations can be made because of the diversity of environmental conditions that shaped their divergences into numerous taxa.

24. With that said, it is useful to do a phylogenetic breakdown of the **eudicots** much as we did previously for general dicot phylogeny. This tree picks up on the node where eudicots branched off.

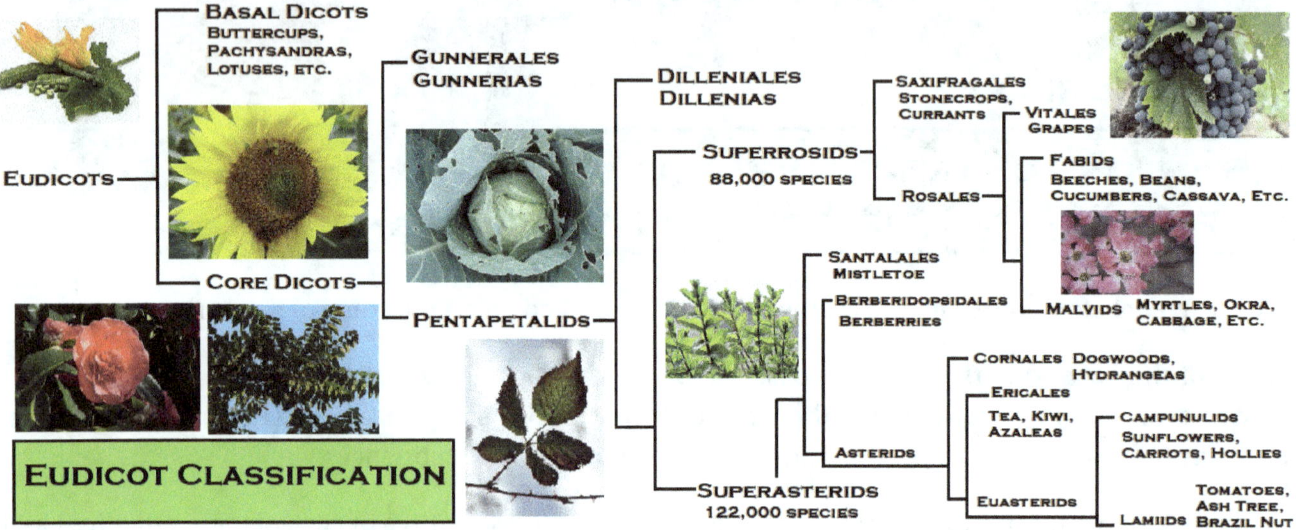

25. The eudicot division of plants is absolutely gigantic. Without expanding into a 1000 page botany encyclopedia, it is impossible to detail every evolutionary direction taken by the quarter million species into their nearly 70 orders.

26. For the purposes of this text, we will concentrate on evolutionary splits and focus on a few of the more prominent orders and families of the eudicots.

27. **Basal eudicots** include a hodgepodge of unrelated groups that don't fit well anywhere else. Examples of the basal dicots include buttercup species, sacred lotus, banksias, and macadamia nut trees.

28. While most of the plants in this **paraphyletic group** are primitive misfits that aren't necessarily close relatives to each other, there are some generalities that tend to mostly be true among the members.

29. First and foremost, the basal dicots have relatively crude and primitive flower anatomy and **monosulcate pollen grains**. Many basal dicot species are also dioecious or have entirely male flowers on certain parts of the plant.

30. They also tend to have a low number of **vessel elements** in their xylem, if they aren't entirely absent.

31. Fossil evidence also suggests that members of these groups were around long ago, and that they veered in different evolutionary paths than the more evolved and more modern **core eudicots**.

32. The diagram below shows some examples of **basal eudicots**.

SACRED LOTUS   MACADAMIA TREE   BANKSIA   OPIUM POPPY

PLANE TREE   HAKEA   BUTTERCUPS   MAHONIA

33. After the split between mainline core dicots and the basal dicots, another couple of divergence events happened. The first was when the **Order Gunnerales** split with the rest of the eudicots.

34. Gunneras have thick succulent stalks and wide leaves. Some species have shockingly large leaves that can be bigger than a dining room table. They have the heaviest petioles of any dicot plant in existence.

35. In addition to having stalks that can be the size of a street lamp, some species of Gunneras have flower stalks that can weigh more than 30 pounds, loaded down with tiny red flowers.

36. Interestingly, many species have symbiotic associations with the cyanobacteria, *Nostoc*, which live symbiotically inside of scales on their stem and help to augment the plant's own leaf photosynthesis.

37. **Order Gunnerales** was split off from the other core dicots, because of DNA sequencing evidence. Besides the symbiotic cyanobacteria, their dimerous (two part flowers) are simpler than most modern dicots.

38. If you are ever in a cloud forest, whether in Costa Rica, Chile, or many other places in Latin America, check out these giant freaks of nature that look like they came from a Dr. Seuss book.

39. The Gunnerales and , as unique as they are, largely got left in the dustbin of natural history. There are only 63 extant species, which pales in comparison to the remaining species of modern **eudicots**. There is a diagram two species below.

GUNNERA TINCTORIA   GUNNERA FRUITS

34. As dicot plants evolved, they became **pentapetalids**, developing distinctly separate **petals** and **sepals**, unlike their more primitive brethren. Flower parts also evolved toward fixed numbers (especially multiples of 5) of organs in their flowers.

35. While some dicots have flower parts in multiples of four (particularly primitive taxa), the repeating pattern of five is so common that this is the derivation of the name **pentapetalid**.

36. The pentapetalids diverged into two points shortly after their evolution. The **Order Dilleniales** veered off into their own directional dead-end of sorts, leaving the rest of the evolving dicots behind.

37. Members of **Order Dilleniales** retain some of the primitive flower structure of the magnolias, lacking defined sepals. They also have uniquely deep secondary leaf veins that go all the way to the margins of their leaves.

38. The flowers of this group have **unilocular** ovaries with **epigynous** pistils, but they are **polycarpellary**, with a single flower producing many different seeds in a cluster of fruit, usually **berries**.

39. Botanists used to think that they were an evolutionary link between magnolias and other more evolved dicots, like violets, based on their physical resemblance, but DNA evidence seems to indicate that they were barking up the wrong tree.

40. For now, no one really knows what to do with them, so they get lumped off to the side. A picture of some of the members of the Order Dilleniales is shown below. You are probably most familiar with the showy peony.

ELEPHANT APPLE   **PEONY**   **HIBBERTIA**

41. Now we get to the real story. Most other eudicots can be parceled into the rose-like dicots or into the aster-like dicots. There are a massive number of orders, families, and species classified into both groups.

42. You will see many examples of related plants that you probably would've never guessed over the next few pages. Take note of the fact that reproductive strategy, reproductive structures, and molecular evidence all carry more weight than morphology.

43. Superficially, the sporophyte (main green part of the plant) provides a lot of suggestive clues about relationships, but similar environments and lifestyles can cause **convergent evolution** to an uncanny degree. Common physical morphologies are often reliable, but it isn't uncommon for similar appearances to lie to you about who is close relatives with who.

44. For instance, some species of cacti of the American Southwest and the spiny euphorbias of Madagascar look nearly identical, but DNA sequences and flowering structures will quickly let you know that they are not at all closely related. See below.

ORGAN PIPE CACTUS
AMERICAN SOUTHWEST

EUPHORBIA INGENS
'COWBOY CACTUS' NOT A
CACTUS AT ALL! MADAGASCAR

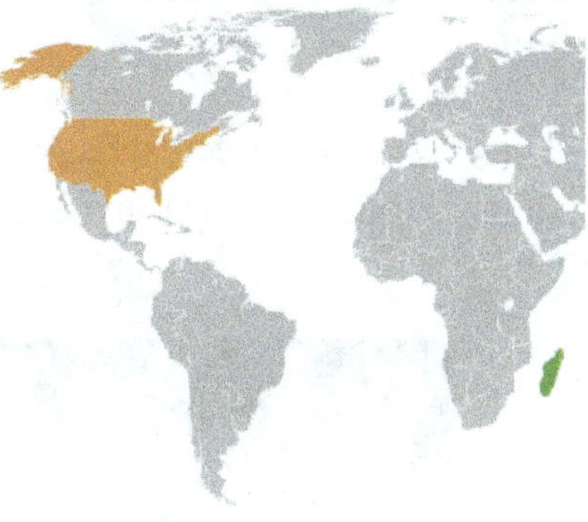

NOT RELATED! FROM
DESERTS ON OPPOSITE
SIDES OF THE WORLD!

45. Let's start off by discussing the delineation between the rose-like dicots and the aster-like dicots.

46. Since almost ALL classification of plants is now done with molecular evidence from genomic DNA, chloroplast DNA, ribosomal RNA, and other molecular evidence, it can be frustrating to understand the logic behind the classification system, since you can't SEE why some groups are being classified together.

47. The **Superrosid clade** unites three different orders. These are the **Orders Saxifragales, Vitales,** and **Rosidae**.

48. There are no reliable uniting physical characteristics that are definitively true for all 88,000 species includes in these three different orders. Only DNA and RNA are foolproof indicators. However, there are noticeable TRENDS that are OFTEN true.

49. For instance, the petals of **Superrosid** flowers are usually not fused, there are usually many more stamens than petals, and their seeds emerge from very large ovules that are held to their fruits by two different **integuments**.

50. It is common for many of the rose-like dicots to have symbiotic nitrogen-fixing bacteria living on or in their roots, and they also often produce secondary compounds in their sap to deter grazing animals and insects.

51. Beyond these characteristics, it is more useful to consider why individual families have been classified together.

52. Let's begin with the two clades that are considered to be members of the **Superrosidae clade**, but that evolved in separate directions from the members of the **Order Rosidae**, the true rosids.

53. **Order Saxifragales** is a dumping ground for plants whose DNA and RNA indicate close relationships, but whose physical features can vary dramatically. Members are as diverse as stonecrops, sweet gums and witch hazel.

54. Again, there are some generalities that are OFTEN true, but there are no ubiquitous defining physical characteristics that are true for all members of the classification.

54. USUALLY, the flowers are radially symmetrical with unfused petals. There are typically **bicarpellary** and **unilocular** carpels surrounded by stamens attached to the base. Their seeds usually develop inside **follicular fruits**.

55. Most members are either woody trees and shrubs, and those that aren't still produce woody secondary growth.

56. The diagram on the next page shows members of several different Saxifragales families. Notice the diversity of plants. Stonecrops are succulents, gooseberries are thorny shrubs, and alders and sweet gums are trees.

57. Members of the **Order Vitales** are also members of the **Superrosid clade**, but not full-fledged members of the true rosids. You know them better as the grape family.

58. Yes, the grape order shares nomenclature with the 'dipsy do dunkaroo they're warming up the bus baby!' college basketball announcer who has obsessively stalked Mike Krzyzewski since the late 1980's. The only thing we really ever learn from Dick Vitale is who currently plays for Duke, along with every miniscule detail of their entire lives. X's and O's? What are those?

59. Members of this group grow as woody vines, have compound flowers that lack conspicuous petals. Instead, they fuse into a green structure called a **calyptra**. Inside, they have a single conical unilocular pistil and five stamens.

60. Mostly pollinated by flies and gnats, the flowers mature into clusters of berries with an array of aromatic phytochemicals and anthocyanins. Typically two to three fruit clusters develop from lateral buds.

61. As compared to most other fruits in nature, grapes have a tremendously high sugar content, which encourages wild animals to feed on them and distribute their seeds in their manure, a ready pile of fertilizer. They also grow in clusters.

62. Most members of this group can readily clone themselves from cuttings via **fragmentation**.

63. Some members of the group, like muscadines, are **dioecious**, but the majority are **monoecious**.

64. Several members of the **Order Vitales** are pictured below. The **Family Vitaceae** is the ONLY family in the order, so all members in the picture belong to the grape family, and are relatively closely related to one another.

MUSCADINE   **COMMON GRAPE**   **VIRGINIA CREEPER**   **CISSUS**

65. Oh you thought we were done with stupid lawsuits? You thought wrong. So sue me....

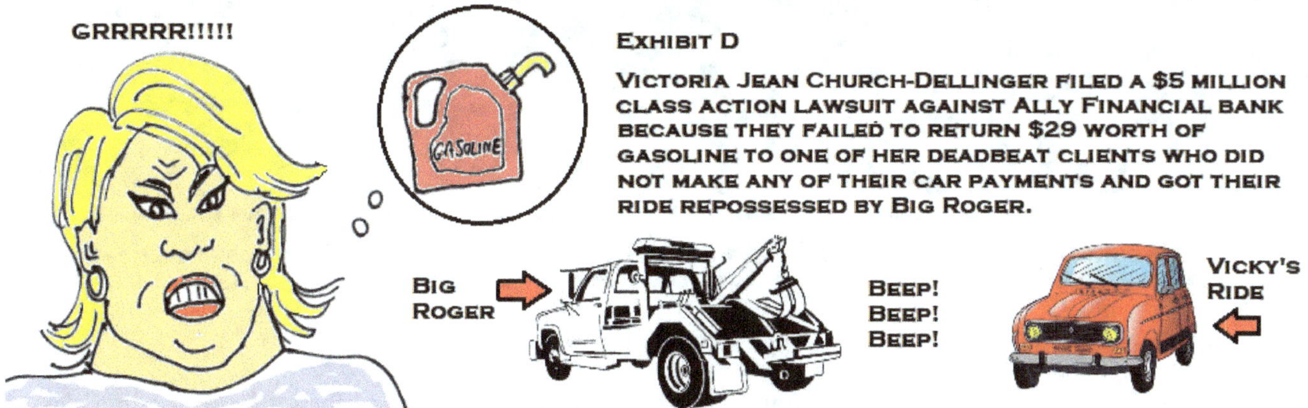

66. Now we get to one of the two largest orders in the entire plant kingdom, the **Order Rosales**. While there is also a great deal of evolutionary diversity in this group, DNA evidence strongly points to common ancestry.

67. There are several morphological characteristics that tend to be seen as a semi-repeating motif within the group.

68. Most members are woody, either growing as vines, shrubs, or small trees. Thorns are common.

69. Most rosids have clusters of little leaf-like growths called **stipules** at the bases of each leaf stalk.

70. Rosid flowers are usually radially symmetrical with flower parts occurring in multiples of four or five. Many members of the group form a **hypanthium** when they bloom, which is a 'cup' of petals.

71. There are numerous stamens in the flower, with the large pistil in the center. There is not an underlying trend,w with regard to fruit development, because members of this group show an array of different fruit types.

72. The rosids are split once again into the **fabids** (bean-like members) and the **malvids** (mallow-like members).

73. The bean-like rosids often have scoop shaped flowers with keel-like petals. They also often host nitrogen-fixing organisms on their roots. Leguminous pod fruits are common among these families.

74. The mallow-like rosids often have cup-shaped flowers with large prominent pistils. Most families within this group make very gummy or mucilaginous compounds in their sap as protection against insects.

75. Some of these physical differences are shown in the diagram below.

GENERALITIES OF FABID-LIKE AND MALVID-LIKE CLADES OF ROSID DICOTS

FABID-LIKE ROSIDS

   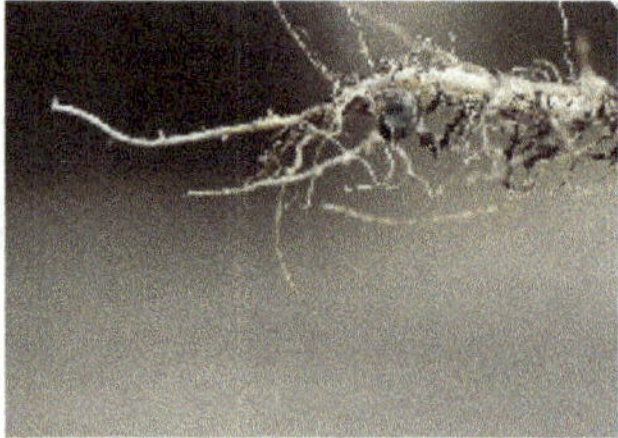

SCOOP-LIKE FLOWERS   **LEGUMINOUS PODS**   **ROUNDED LEAF EDGES**   **ROOT NODULES**

MALVID-LIKE ROSIDS

CUP-SHAPED FLOWERS   **CAPSULES**   **SERRATED / DIVIDED LEAF EDGE**   **BERRIES OR POMES**

NOTE: NONE OF THESE TRAITS ARE TRUE FOR ALL OF THESE CLADES, THESE ARE JUST ONES THAT ARE SEEN AS A SOMEWHAT COMMON MOTIF AMONG MANY DIFFERENT FAMILIES OF PLANTS.

75. Let's begin with a summary of several of the major orders of **fabids**.

76. We will not cover them all, as there is not much to be gained but the need for a pair of reading glasses and knowledge of obscure plants that only about 33 people (plant taxonomists no doubt) on the planet care passionately about.

77. The **Order Malpighiales** includes a huge diversity of plants. Many seem like they couldn't possibly be related, but DNA says otherwise, speaking to the adaptability of the plant gene pool. They do have some general physical commonalities.

79. Similar design in flower form provides clues that unite the group. Their flowers typically have 4 or 5 petals, and rather than nectaries, they mostly give pollinating bees wax.

80. Their flowers are **tricarpellary** or **polycarpellary** with **hypogynous** fused pistils. Their numerous single anthers (which can be 10 or more like many of the more primitive basal dicots) are often **basifixed.**

81. Members of this group typically have **opposite toothed leaves** with **stipules** at the base.

82. There is a great amount of physical diversity in the group and the order contains everything from toxic spiny euphorbias to weeping willows to passionfruit.

83. The Malpighiales have plenty of members that also specialize in being nasty to herbivores.

84. Castor beans contain the toxin ricin, which jams up ribosomes and is one of the most toxic substances known to man. A single castor bean will kill a full grown man. Animals with any common sense avoid them.

85. Many of the euphorbias from Madagascar hit you with a double-whammy of thorns an extremely caustic latex sap, full of turpene esters. Plants like desert rose and pencil cactus (not really a cactus) can give chemical burns.

86. Willows produce salicylic acid in the phloem under their bark to discourage insects from eating them. This hasn't deterred people from harvesting the bark, since this is one of the starting ingredients for aspirin.

87. Poinsettias, a favorite of decorating moms at Christmas contain toxic latex. At best, they will give chemical burns to the mouths of stupid cats. At worst, they will send Mr. Jingles to the big litter box in the sky.

88. Molecules of some of these Malpighiales toxins are shown below.

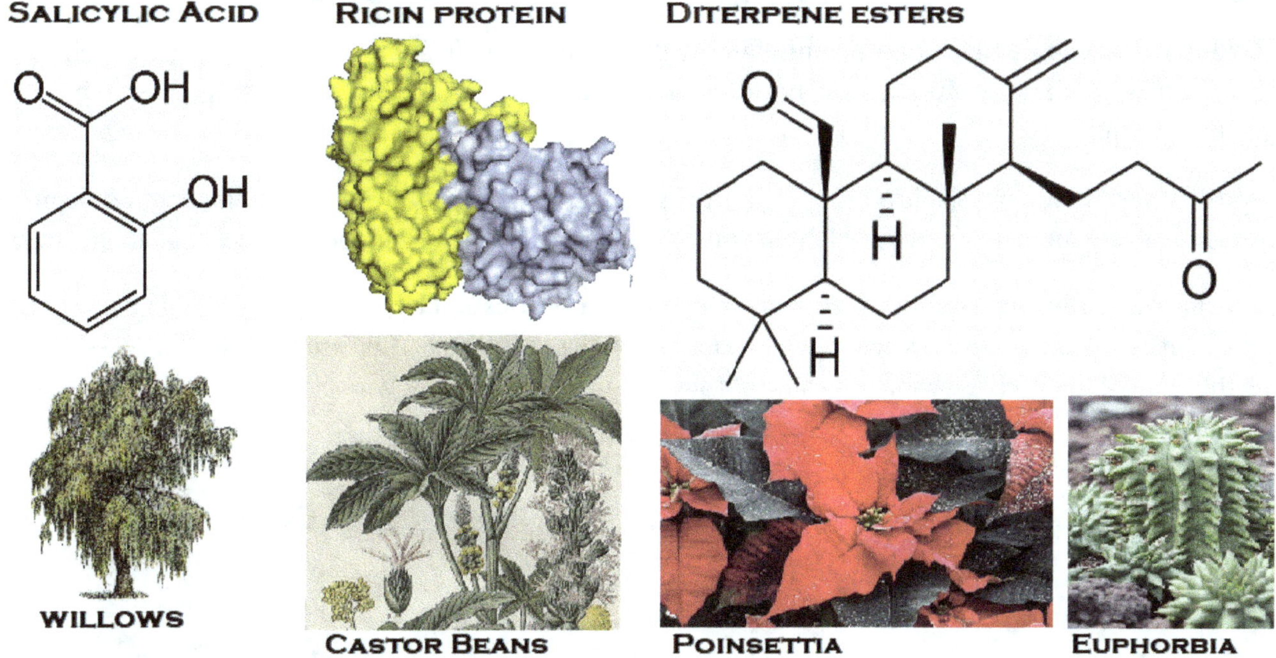

89. However, some members of this group haven't gotten the message that they are supposed to be toxic.

90. Passionfruit is a food staple in the tropics, while the British love their violet-flavored candies.

91. This group is a favorite of evolutionary biologists, since the plants of Madagascar are kind of like a botanical archetype for the marsupials of Australia. Euphorbias underwent an **adaptive radiation** there.

92. There are euphorbias that are closely related, according to DNA, that would never be recognized as plants from the same order, let alone family. Waxy shrubs growing alongside cactus-like giants are often related.

93. The diagram below contains a woefully inadequate cross-section of some of the many species in this order.

94. **Order Fabaceae** contains one gigantic family, the bean family (legumes) and several other relatively obscure families. **Nitrogen-fixing root nodules** and **symbiotic** associations with nitrogen cycle bacteria define this group.

95. Pod-like seed cases and shallow drooping scoop-shaped flowers are typical of its members. While many of its members are annual or perennial herbaceous plants, there are also a number of trees that are legumes.

96. Legumes have **monocarpellary** flowers with **unilocular** ovaries and **hypogynous** pistils. The flower stamens are **bithecal**. These ovaries mature into bean pods with seeds (ovules) that are **marginally placentated** along the edge of the pod. A diagram of some of these commonalities is shown below.

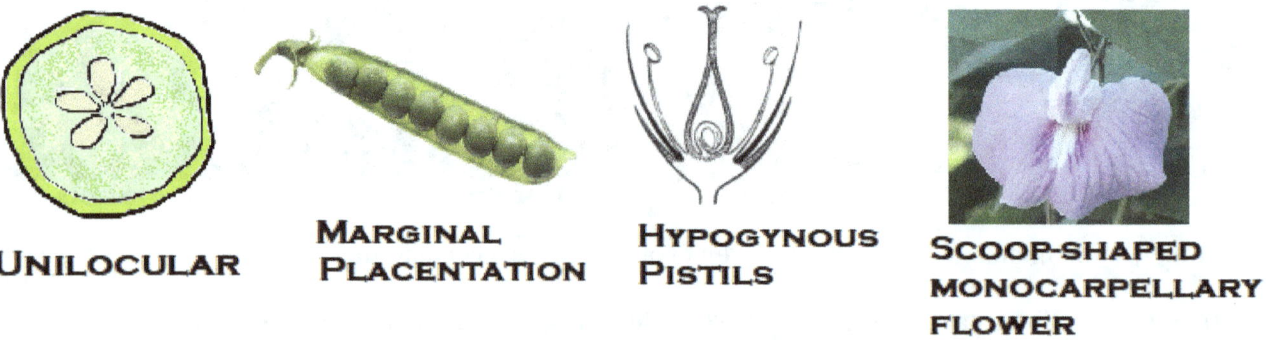

97. Green beans, sweet peas, lima beans, black-eyed peas, navy beans, pinto beans, greasy beans, wax beans, beam beans, chickpeas, snap peas, snow peas, and lentils are all legumes. Also...peanuts are really beans too.

98. There are many tropical trees and a number of temperate species that are legumes. Their flowers and fruits are giveaways that they are relatives of the beans, since they produce a row of seeds enclosed in a pod.

99. African acacia trees are a food source for giraffes, baboons and antelope. Royal poncianas are ecologically similar in the tropical new world for monkeys and grazing animals. Mimosa and locust pods are eaten by squirrels, pigs, and other grazing animals in temperate habitats.

100. Almost all legumes have nitrogen-fixing and nitrification bacteria living in **root nodules** symbiotically. In exchange for a ready source of nitrates for the production of proteins and nucleic acids, the bacteria are given sugars and a sheltered home.

101. The diagram below shows a cross-section of some of the members of the Order Fabaceae.

102. The **Order Rosales** contains its namesake roses, as apples, peaches, blackberries, elm trees, strawberries, almonds, figs, nettles, hops, and the plant that will get you a free trip to jail if cops find it in your car.

103. Members of this order are either woody shrubs or trees or capable of woody secondary growth, if they herbaceous. Almost all members have alternate leaf arrangement with **stipules** at the base.

104. Flowers are typically shallow and cup shaped, with lots of stamens. Their pistils are **perigynous** and their ovaries are **unilocular.** Many rosid flowers have numerous fused pistils in the middle. Most are heavily pollinated by bees and butterflies. Typically, colored petals and green sepals form separate distinct layers underneath the reproductive parts of the flower.

105. Members of the rose order have **perfect flowers**, meaning that all male and female parts are present within the same flower. The **stigma** is usually dry, with nectaries at the base of the stamens instead. The **sepals** usually fuse at the base and the flowers are shaped like cups.

106. The fruit of Rosales are often fleshy **drupes** (stone fruits), such as cherries, peaches, or plums. **Achenes** are also common, where a dry fruit fuses itself to the seed coat, such as in rose hips and strawberries (the true

fruit is the case on the seeds, what we call a fruit is a swollen stem). **Aggregate fruits** of both types are also common, such as in blackberries and raspberries.

107. The diagram below shows some of the common morphological traits of rosids.

108. While they aren't as prevalent in agriculture as monocot grasses, nevertheless, members of this order comprise a tremendous number of agricultural species, particularly important to world fruit and honey production.

109. The diagram below shows a few of the hundreds of diverse species that make up this group of dicots.

110. A large number of dicot tree species belong to the **Order Fagacae**. Like their legume relatives, root microorganisms are critical to the success of its species, although in this case, it's fungal **mycorrhizae** they need, rather than bacterial symbiotes.

111. Root fungi greatly expand the surface area of the tree's roots, allowing it to take up adequate water and soluble minerals. In exchange, the mycorrhizal fungi get a cut of the sugars produced by the photosynthesizing tree.

112. Mostly without exeption, all members of Order Fagacae are woody shrubs and trees. They produce separate **imperfect flowers**, with brush-like male **catkins** and simple female flowers on the same hermaphroditic tree.

113. Most members of the Fagaceae produce dry fruits that are familiarly called **nuts**. The shell (such as pecans, hickories, acorn caps) is actually a dry fruit (former ovary), with the seed (former ovule) is the nut inside.

114. Leaves are diverse in form, but depending on speices, are commonly lobed or lance shaped, sometimes pinnate, and typically have **stipules** at the base of each **petiole**.

115. Most of the trees in this group are found in temperate climates and have adapted to the colder winters by becoming **deciduous** and living off of stored sugars in the roots and sap when solar energy is low.

116. Some common traits of Order Fagacae are shown below.

118. Economically, this group is important for tree nuts and for high grade wood. Pecans, walnuts, hickories, birches, oaks, and beeches all belong to this order. You can't walk through a temperate forest without seeing this group.

119. Often these species are involved as kindling for forest fires because wood is flammable. Okay that's a very poor transition, but it's an excuse to present the next profile of a shockingly dumb lawsuit.

120. Fire bad. Me file lawsuit.

120. The diagram below gives examples of a few prominent tree species that represent Order Fagacae.

121. Now we move on from the forest to the garden. The **Order Cucurbitales** includes the familiar pumpkins, squash, melons, and gourds of the vegetable garden, as well as the begonias of the flower garden.

122. Many members of the Cucurbitales make protective aromatic compounds in their leaves, stems, and fruits to prevent consumption by slugs, insects and parasitic fungi. Some of these compounds called **cucurmins** are strong anti-oxidants that reduce inflammation and may even prevent cancer.

123. The Cucurbitales are often **monoecious** or **partially dioecious** with separate sexes or separate sex flowers on the same plant. Anyone who has grown squash knows that you have to have male and female flowers near one another to get any squish squash. Squish squash eat your applesauce. The flowers themselves have **epigynous** pistils, cone-shaped anthers and five petals.

124. The **trilocular ovaries** of the fruit grow into three chambers as they develop into giant fleshy berry fruits called **pepos.** One hallmark of pepos is that they have **peripheral placentation**, with seeds anchoring their funiculi toward the outside.

125. The leaves of Cucurbitales often have deep lobes, while the stems have distinctive **vascular bundles**, with the xylem tissue sandwiched between two layers of phloem. Trichomes are common. The diagram below shows some common cucurbit traits.

126. While it is more common for the members of this group to be herbaceous plants and vines, there are many tropical trees and woody vines that are members of this order as well. Most have long carrot-like **taproots** with lateral side roots.

127. The diagram below shows some representative members of the curcurbits.

128. Now we have reached the point of the great rosid split. We will now move from the bean-like rosid eudicots to the mallow-like eudicots. Members of the **Malvid clade** represent more than 25% of all dicot species.

129. While DNA sequencing tightly lines up the organization of their taxonomy, physical characteristics are varied. Almost all malvids have cup-like flowers and produce gooey mucilaginous compounds in their sap.

130. With at least 17 known orders of these plants, we will stick to a few of the most prominent. Malvids include citrus, cashews, mallows like okra and cotton, myrtles, geraniums, mahogany trees, and maples.

131. **Order Geraniales** includes mostly herbaceous annual and perennial wildflowers. They have generic **polycarpellary** perfect flowers with ten or more stamens. The five petals are symmetrical. There are alternating whorls of stamens inside the flower, with **nectaries** at their base. They mature into **capsule** or **schizocarp** dry fruits.

132. Leaf arrangement in this order is spiral or opposite on a fleshy stem, though some species are woody shrubs.

133. With a few exceptions in the flower garden, creosote for telephone poles, and one BIG exception from South America, most of the members of this group are not very important commercially.

134. However, coca is sadly very valuable, you could say. It drives the economies of Colombia and Peru, with an unbelievable production of nearly 2,000 metric tons per year, with a street value of $150 billion. Worse, nearly 100,000 civilians per year die as a result of drug wars in Mexico, Colombia, Venezuela, the USA, and other major trafficking hubs.

135. Besides these few exceptions, most of the members of this group are mostly obscure weeds and shrubs. Some examples of various members of this order are pictured below.

136. Members of the **Order Myrtales** include the myrtles, eucalyptus, primroses, leadwoods, and pomegranates. Flower structure varies across taxa. Some have epignyous flowers, while others are perigynous. Flower parts are usually in fours and fives and the stamens are often clustered together in **didynamous** arrangements, though they are greatly reduced in number in some groups.

137. A distinctive characteristic of the members of this order, is that the wood **xylem** is sandwiched with two layers of **phloem**, rather than having the characteristic single layer under the bark in most dicots.

138. The leaves of many members of this order are dotted with small **oil glands** that produce aromatic secondary compounds. Examples of this include the bay leaves used in Cajun cooking and eucalyptus oil used in cough drops.

139. While there are some exceptions, most members of Order Myrtales are woody trees or shrubs with smooth bark that produce fruit as berries. Many groups make repellant secondary compounds, such as the koala-favored eucalyptuses.

140. The seeds of members of this order often lack an **endosperm** and solely rely on **cotyledons** to nourish their embryos.

141. The primroses are an interesting family within this order, as there are both **diploid** and various **polyploid** species. Some of hybridizations between diploid species can serve as a mechanism to form new species with multiple chromosomes.

142. Root modifications are common among the Order Myrtales. Many families, such as eucalyptus produce small bud-like pockets of meristamtic tissue called **lignotubers** near the base of their trunk. In the event of fire, these pockets of tissue allow the tree to rapidly re-grow branches and leaves.

143. Mangroves, which are circumglobal in the Tropics, have modified their roots into **prop roots** covered in **lenticels**. This adaptation prevents the trees from suffocating in the low oxygen still waters of the swamps where they live.

144. Some members of the Order Myrtales are shown below.

145. Members of the **Order Sapindales** are well known to everyone. Citrus, mangoes, cashews, maples, mahogany wood, frankincense and myrrh bushes, lychee fruit, sumac, buckeyes, and poison ivy all belong to this order.

146. With only a few exceptions, members of this group are woody trees, shrubs, or vines. Compound leaves are the norm, with leaflets arranged alternately on a central petiole. Many species also modify leaves into thorns.

147. Citrus trees seem to have de-evolved the petiole into a joint that joins the main leaf to the stem. Instead, they have a jointed leaf that dips in the middle, with a small lower lobe and large uper lobe.

148. Flowers are **monocarpellary**, but the ovaries themselves usually have multiple **locules**, resulting in fused **capsule** fruits in many types of Sapindales, while in citrus, a modified segmented berry called a **hesperidium** results. Inside citrus fruit, the seeds are centered, but they are not anchored in the center. Rather, they have **axillary placentation.**

149. Flowers produce copious amounts of **nectar** from glands at the base of the flower between the stamens and pistil in a disc-like channel. Most members of the order have fragrant blooms to attract insect pollinators. Orange blossom honey, for instance, is one of the most widely produced types of honey in apiaries around the world.

150. Overall, fruit types vary tremendously between different families. Citrus produce the aforementioned **hesperidiums,** sumacs produce **berries,** maples produce winged **samaras,** lychees are **nuts,** and mangos are stonefruit **drupes.**

151. Most members of the group are at least partially **dioecious**, so many evolved protections against self-pollination. Often, these plants default the sex of each flower by whichever gender parts develop first, rendering the undeveloped organs sterile. Others have genetic mechanisms to prevent pollen from fertilizing their own ova.

152. Another common occurrence in this taxa of plants is the presence of **oil glands** throughout the sporophyte.

153. Some of these oils, like urushiol in cashews and poison ivy are highly irritable. Citrus trees also secrete bitter oils as a protection against insect consumption. Frankincense and myrrh have essential oils valued for their smell.

154. The diagram below shows some of the botanical features found in members of the Order Sapindales.

155. The diagram below contains representative members of the Order Sapindales.

156. **Order Malvales** lends their name to the entire clade of malvids. These plants are commonly known as mallows.

157. Numerous well-known plants are types of mallows. These include okra, cotton, hibiscus, cacao, kapok, baobob, daphne, hollyhocks, kola nuts, durian fruit, Rose of Sharon, and even the marshmallow.

158. Yes, the marshmallow is real. This isn't a snipe hunt. The marshmallow was the flavoring agent for marshmallows in the early days, before food companies went with artificial flavorings.

159. So what makes something a mallow? One uniting characteristic is that these plants are slimers. If you have had to watch old people eat slimy boiled okra, you know. These plants make lots of booger-like mucilaginous slime in their tissues.

160. The leaves of mallows are usually palmately lobed and also have deep palmate veins. Prickly **trichomes** (leaf hairs) are also very common. You also know this if you have had to pick okra and felt like you grabbed a handful of fiberglass.

161. The flowers of mallows are **hypogynous** and in parts of five, with a fused **calyx** base called a bract that firmly holds the trumpet-shaped flowers onto the plant. The ovaries are **pentalocular**, which can be clearly seen in the 5 segments of an okra pod. Some members of the order are **partially dioecious** with separate sex flowers.

162. Regardless of whether the stamens and pistils are on the same flower, the anthers are **monothecal** and **basifixed.**

163. Mallows can be herbaceous with woody secondary growth or full-fledged bushes and trees, depending on species. Bark is usually thin and fibrous with a green phloem layer right under the surface.

164. Almost all mallows are extremely fibrous in all parts, with lots of **collenchyma**. For example, the stalks of cotton and okra are loaded with cellulose and nearly impossible to pull apart by hand. Cotton, itself, is pure cellulose fibers.

165. The most common type of fruit produced by mallows are dry **capsules**, such as okra or cotton bolls. However, cacao produces fleshy **pepos**, while durian fruit has several sections carpels, so it is a **hesperidium.**

166. The diagram below shows several common features seen in the mallows.

FLOWER PARTS IN MULTIPLES OF 5.

PENTALOCULAR OVARIES (DURIAN IS A HESPERIDIUM)

FIBROUS PLANTS; LOTS OF COLLENCHYMA

BASIFIXED MONOTHECAL ANTHERS

PENTALOCULAR OVARIES

CACAO IS A PEPO

MUCILAGINOUS SLIME

PINNATE DIVIDED LEAF

167. Durian fruit, the most beloved mallow in the Orient, is known to smell like a dirty diaper baking in a sunny window of a public bus station on a hot August day. It apparently tastes like a mix of a mixture of garlic, old cheese, and whipped cream.

168. Its texture is said to be like slimy melted brie cheese. It is banned from subways in many Asian countries, because the smell can clear out an entire car. Now for a quiz about everyone's favorite Malvid....

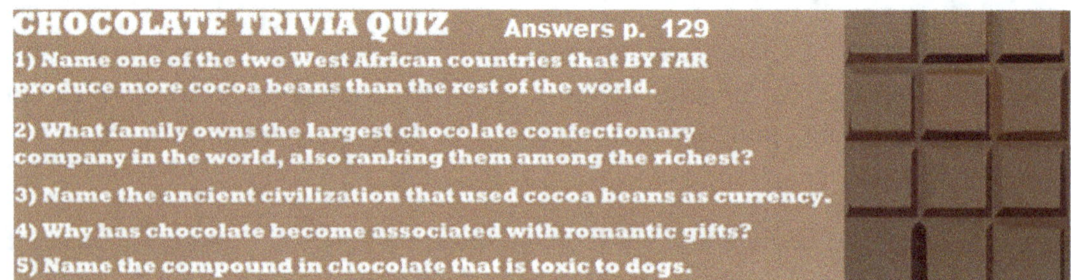

169. The diagram above shows some representatives of the Order Malvidae.

170. The **Order Brassicales** contains mostly herbaceous plants that grow in cool temperate regions. They are the bane of children and enlightened adults everywhere, as they give us kale, Brussels sprouts, cabbage, turnips, cauliflower broccoli and other gross vegetables that have to be eaten to earn dessert.

171. With that said, brassicas are a diverse order, containing many tree species, mignonette herbs like rocket, nasturtiums, papayas, capers, and a boatload of obscure herbaceous weeds.

172. Members of **Order Brassicales** make protective sulfurous compounds, such as allyl-isothiocyanate, hence the stank of cooking broccoli. Mustard oil is potent enough to kill insects that try to eat the plant and cause chemical burns in concentration.

173. Typically plants in this group make **racemose** flower stalks, meaning that when flower buds grow laterally, the main stalk keeps growing and putting out new flowers at the tips. This results in large clusters of flowers.

174. Brassica flowers have **tetralocular hypogynous** pistils. Two ovaries fuse at their base, in the center, so their flowers are **bicarpellary**. The anthers are **basifixeed** and **tetradynamous**.

175. The flowers of plants in this group are also unusual, because the nectar glands are located between the petals and stamens, rather than in between the stamens and the pistil, which is the usual arrangement.

176. The seeds of many of the brassicas are among the smallest known in the plant kingdom, hence the Bible parable about having faith as small as a mustard seed allowing one to move mountains.

177. The seeds, themselves, usually have green embroys and grow in pod fruits that split open on the sides when dry to re-seed the plant. They are **marginally placentated**. Since many species are annuals and/or invasive weeds, they are **R-strategists** and may produce thousands of seeds, with the hope that someone survives. They are rich in oil, rather than carbohydrates.

178. **Infloresence** varies between the herbaceous brassicas and the ones that develop into woody trees and shrubs.

179. The diagram below depicts some of the common traits seen in Brassicas.

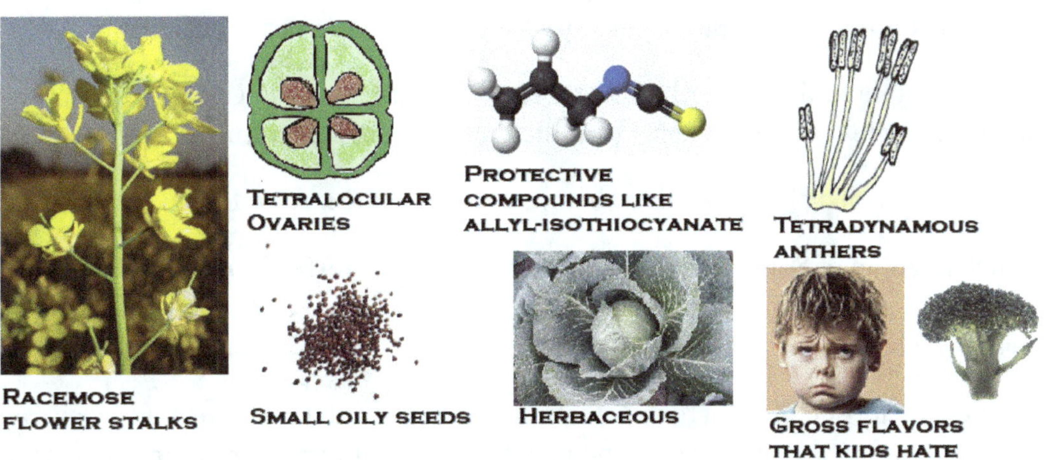

180. The picture below depicts some of the many members of Order Brassicaceae.

181. Now we return to the base of the modern eudicot tree and veer down the other branch. We will now discuss the aster-like superclade, the **Superasterids**, which have fundamental anatomical differences with the **Superrosids**.

182. Like the Superrosid superclade, there are several groups of more primitive plants that evolved in different directions earlier than the more-evolved mainline **Asterid clade**.

183. The most primitive order of aster-like plants are found in the **Order Santalales**, which includes the sandalwood trees and mistletoe. They are rudimentary for several reasons, as compared to most flowering plants.

184. First, members of the Santalales have lost their **seed coats**. Second, their leaves have lower concentrations of chlorophyll than most plants, leading some members to become partial or obligate parasites on other plants.

185. With the exception of the sandalwoods, almost every other member of this order lives by attaching themselves to the stems or roots of other plants and stealing sugars and nutrients from the phloem.

186. Some plants like witches broom and Balanophora lack chlorophyll entirely and could be confused for a parasitic fungus to an untrained eye. Both plants sponge off of living tissues in host plants.

187. Flowers are hermaphroditic or start hermaphroditic and terminate the gender parts that are slower to develop, becoming a single sex. Many members of the group produce their fruit as berries.

188. Members of Order Santanales are pictured below, along with the next order of plants.

189. Another ancient group of **Superasterids** belong to the **Order Berberidopsales**. This order contains a handful of small desert shrubs native to the Southern hemisphere. These are commonly called coral plants. One of these is shown below.

190. One unique characteristic of this group is that their stems or trunks are covered by a scaly epidermis. Their leaves are in opposite arrangement. Some members of the group are **monoecious**, while others are **dioecious**.

191. Without molecular evidence from DNA sequences of the nucleus and chloroplast, however, there aren't enough satisfying obvious differences between these and mainline eudicots to split them off.

192. DNA evidence indicates that a second evolutionary divergence occurred, splitting the Superasterids again into the **Order Caryophyllales** and into a second group that evolved into 17 orders in the mainline **Asterid clade**.

193. The **Order Caryophyllales** is an enormous classification with 37 families and more than 12,000 species. It is quite possible to see two members of this order and assume that they are nowhere close to being related. Most commonalities are genetic in nature, so it is hard to make a lot of generalities based on morphology.

194. However, besides clear DNA evidence linking their evolutionary relationship with one another, their metabolism also gives a big clue to their relatedness. Many of the plants in this group use **C4 or CAM photosynthesis**. The diagram below shows how the C4 pathway diverges from the normal C3 photosynthetic pathway.

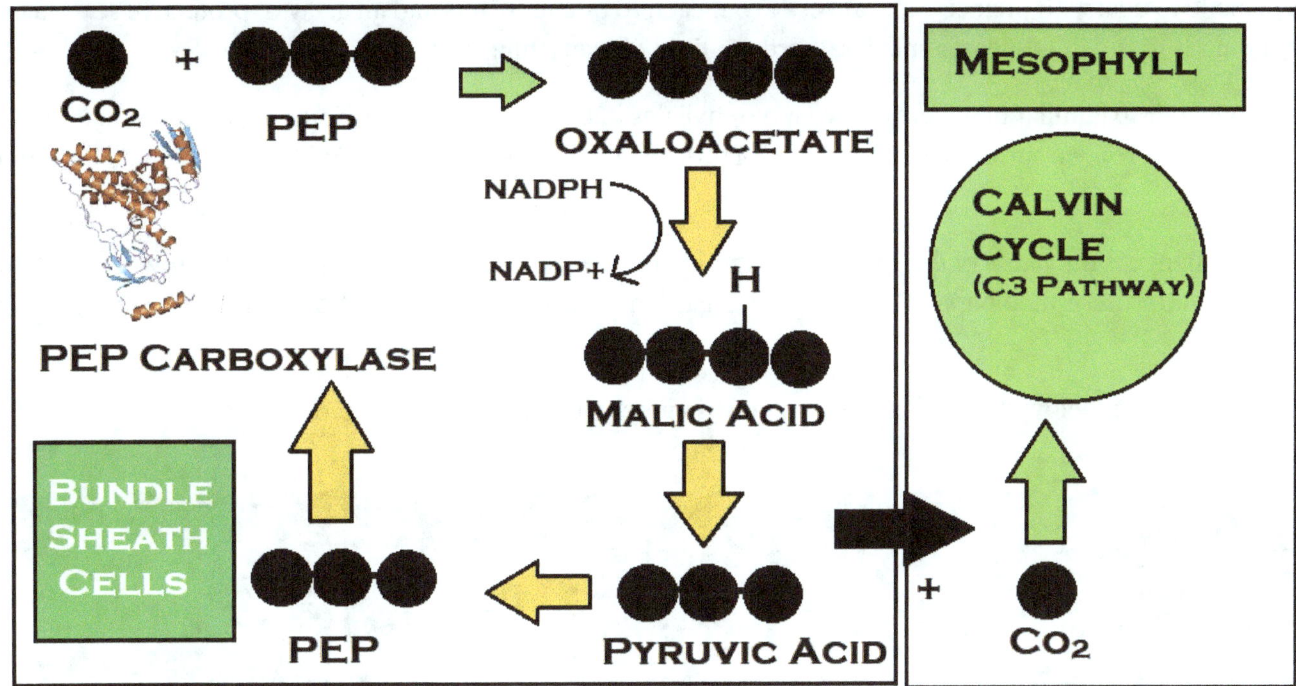

195. Plants in this group often have swollen **bundle sheath cells** for storage of **malic acid (C4)** and they use the enzyme **PEP carboxylase** to fix carbon, so that it can be stored without having to open the **stomata**. The difference in C4 leaf anatomy is shown in the diagram below.

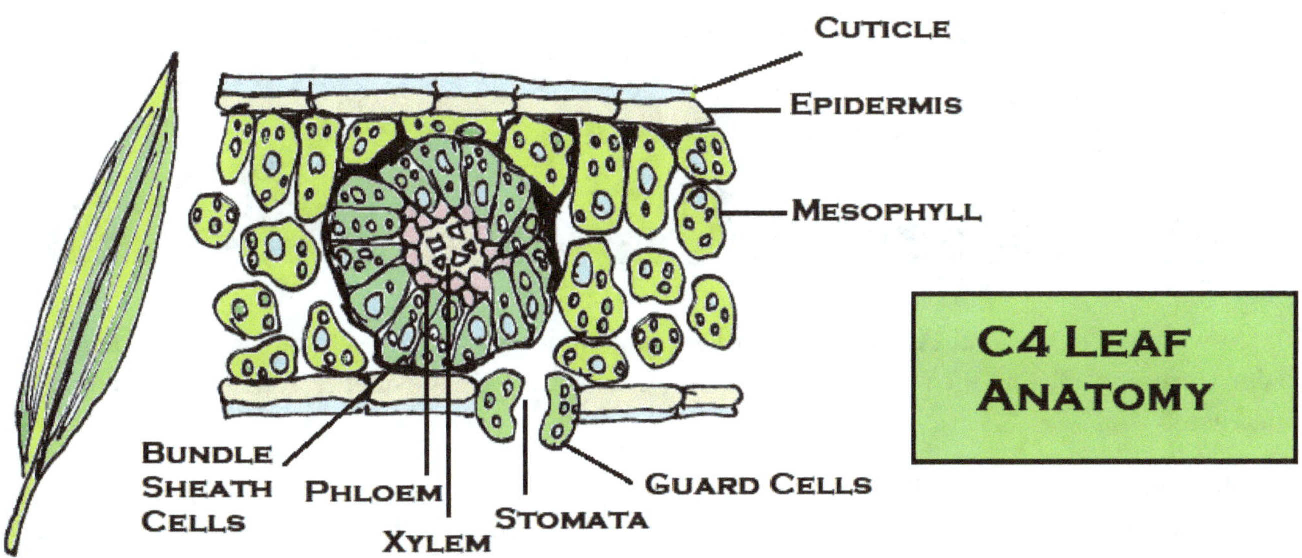

196. C4 metabolism lets tropical and desert members of this order keep running the Calvin cycle and producing sugars, without having to take the risk of opening the stomata and dehydrating. Most members off the group also have an abundance of red and purple **anthocyanin** pigments to harvest as much light as possible to run their energy-demanding C4 pathways.

197. Perhaps due to the relative rarity of soil fungi in arid desert climates, many members of this order lack root **mychorrhizae.**

198. One hallmark of this interesting order of plants it the wide array of weird evolutionary adaptations they have made to deal with their circumstances. From cacti to pitcher plants and an array of protective secondary compounds found in numerous species, there is a little of everything in this diverse group.

199. Some of the familiar members of the Caryophyllales include the cacti, many other succulents, marigolds, carnations, portulaca, insectivorous plants, sorrels, buckwheat, smartweeds, pokeweeds, purselanes, and amaranths.

200. Resilience is the name of the game in this group. From nasty thorns to water storage to growth rates that make some of these members the most noxious garden weeds known, most of these plants are tough as a bawse.

201. The diagram below shows some of the many members of this gigantic group of plants.

202. Now we get to the **True Asterid clade,** wherein another evolutionary split occurs that separates the **Order Cornales** off from the rest of the aster-like plants, due to DNA evidence and some morphological differences.

203. Plants in this order are distinct, because their flower parts occur mostly in multiples of four. Almost all members of this group are woody shrubs or small trees and their fruits are **drupes** with very large hard single seeds.

204. Dogwoods, and many other members of this order, develop a compound flower with many tiny individual flowers, sounding this cluster with modified leaves called **bracts** that are mistaken for petals by most.

205. Other members of the group, such as hydrangeas, develop huge sphere-like clusters of many smaller flowers without using bracts to advertise their presence. **Anthocyanins** are common pigments in these flowers.

206. Since anthocyanins are acid/base indicators, the color of these pigments can change according to soil chemistry. Many types of hydrangeas grown in alkaline soil have blue flowers, while pink flowers indicates acidic soil.

207. **Order Cornales** includes the temperate dogwood and tupelo trees, the tropical African lancewoods and sillyberry trees, shrubs like hydrangeas, and the herbaceous stickleaf weeds of the Southern hemisphere.

208. Members of this order have distinct four-part flowers, rather than the more typical multiples of five. A central flower develops first in the cluster from a **terminal bud**, later producing more flowers from lateral offshoots. They produce **drupe** fruits.

209. The picture below shows several members of the Order Cornales.

210. Now we get to the last group of plants that doesn't 'make the cut' to get membership into the most evolved group of aster-like plants, the **Euasterid clade**.

211. The **Euasterids** have flowers with a single whorl of stamens and flowers with bilateral symmetry, while members of the Order Cornales and the **Order Ericales** have multiple stamens and sometimes radial flower symmetry.

212. The **Euasterids** are also specialists at manufacturing protective **alkaloid** chemicals like nicotine, solanine, and hundreds of other toxins, which are not commonly found in the more primitive Cornales and Ericales

213. Besides not making it into the club for this reason, members of these 'not quite good enough' plants also don't necessary have the **bilocular** (two chambered) ovaries or the single whorl of stamens found in **Euasterid** flowers.

214. DNA evidence, which is also not very gratifying to people who want to SEE what the differences between groups, also pegs the Order Ericales as an older group of plants.

215. Don't worry though....this whole classification thing will probably get scrapped and re-done again in 10 years anyway, because one thing that taxonomists love to do is argue about obscure stuff that regular people like plumbers and janitors don't care about.

216. At some point, there has probably been a parking lot fight between two biology professors arguing over how to classify a new slime mold into a genus. The fight was probably almost as sad as an NBA 'fight' where two guys have to be 'held back'.

217. All that really ever happens in an NBA 'fight' is that two seven foot tall guys pretend to throw punches after a foul, never actually connecting, and then they dance around waiting on teammates who 'barely' pull them apart.

218. The last real fights in the NBA went away when Larry Bird, Bill Laimbeer, Rick Mahorn, and almost the entire Trailblazers roster of the 1990s retired. Don't even bring up Ron Artest......I'm sorry....I mean Metta World Peace, the 250 pound 6-8 pro athlete who punched a fat, balding seated fan in the face while he was eating.

219. Okay, back to the **Order Ericales**. One reason Ericales don't make it to into the **Euasterids** because their flowers tend to have fused petals and they have more than one whorl of stamens. Both are considered to be older traits.

220. However, it is still DNA evidence that provides the strongest reasoning.

221. The **Order Ericales** includes azaleas, blueberries, huckleberries, persimmons, Eric Clapton, Erik Estrada, kiwi fruit, tea, rhododendrons, mountain laurels, Eric Bienemy, Eric the Red, camellias, phlox, Eric Dickerson, and Brazil nuts.

222. Berries are the most common type of fruit produced by members of this order, but fleshy drupes and dry capsules of various types and sizes are also pretty typical.

223. In the case of the Brazil nut, these can be cannonball-sized hard fruits that can kill someone when they drop if they happen to be standing in the wrong place at the wrong time.

224. Members of the **Order Ericales** are pictured below.

225. Now we finally embark on a journey through the true **Euasterids**, which is a gigantic taxa of plants that includes 15 different orders of plants, spanning some 60,000 different species.

226. It is thought that the **Euasterids** evolved roughly in sync with the emergence and evolution of insects.

227. Thanks to **coevolution**, a common theme among this group is flowers that fit their particular species of insect pollinators like a glove. Each member of the **mutualistic** relationship adapted to the needs of the other.

228. Most members of the Euasterlid clade have stamens that project well above the petals, which often sit in a flat plane like a dinner plate. Their nectar glands are fused into a disc at the base of the flower.

229. Most members of this group have perfect flowers with parts in multiples of four or five. While the types of fruit made by Euasterids run the gamut of types, the seeds are typical held onto the fruit by a single **integument**.

230. The Euasterids are further divided into two more sub-clades called the **Campanulids** and the **Lamiids**.

231. These sub-divisions mostly have to do with the appearance of the developing flower as it first emerges from the germ cells that produce them.

232. **Campanulid** flower germ cells look like a flat disc a la hockey puck. As the flowers develop, the middle caves in, forming a star shape. Petals form from the cells on the side, while stamens and pistils form in the middle.

233. **Lamiid** flower germ cells are not in one flat layer. Instead, there are lump-like balls of tissue that will form petals and stamens sticking up in two concentric rings. One has to use a powerful microscope to see either germ line.

234. With that said, most of the reasoning for sub-classifying these two groups into sub-clades is based on DNA evidence. We will spare any further hair-splitting and directly discuss a few of the most prominent orders of both.

235. Let's start with the **Campanulid** sub-clade. We will only cover a few of the most well-known groups.

236. The namesake group of the aster-like dicots is the **Order Asterales**. These include the familiar asters, daisies, sunflowers, and bellflowers, as well as hundreds of species of herbs, many shrubs, and a handful of trees.

237. Members of this order have plunger-like groups of fused stamens, which form a hollow tube in the middle. The tube then fills up with pollen that drops from each anther in the circle.

238. Typically, these specialized pollen plungers are located right next to the female style, ensuring that the fluffy rear ends of visiting bumblebees or honeybees will fertilize the flower, while also picking up more pollen.

239. The flower structure of members of this order are usually compound, with dozens or hundreds of small flowers combining to form a large flowerhead, like a sunflower or zinnia. The outermost flowers grow petals that join in a ring-like corolla. What appears to be a single flower is actually numerous individual flowers.

240. The pistils of asters are also capable of storing back pollen inside the style for later.

241. Biochemically, one odd trait of the asters is that they don't store starch in their seeds or the cortex tissue of their stems and roots. Instead, they produce and store another complex carbohydrate called **inulin**.

242. The fruits of most asterales are **achenes** with **basal placentation.** Sunflowers are a good illustration of this, as they have dozens to hundreds of seeds attached at the

243. Members of **Order Asterales** are represented in the diagram below.

244. The **Order Apiales** is a favorite of chefs and Bugs Bunny. It is the taxa that includes carrots, celery, parsley, and numerous other herbs. Ginseng, mock orange, aralias, and a number of tropical shrubs also belong to this taxa.

245. Plants in this order are characterized by bottle-shaped pistils and flower ovaries that have two lobes. These often mature into a dry fruit that splits in half to release the seed. Other species produce fleshy drupes.

246. The flowers themselves are **perfect flowers** that often grow in compound clusters. Their petals are claw-like.

247. Most Apiales produce aromatic compounds as protection against insects, fungi, or bacteria. This has the unintended effect of people finding many species delicious, and picking them and using them for spices.

248. Many of the members of the Apiales are short-lived herbs with either **annual** or **biennial** life cycles. For instance, carrots spend their first season storing food (the carrot) and use the energy to go to seed in year two.

249. The Apiales are more commonly found in cooler temperate habitats, as they are **C3 plants**. The diagram below represents several members of the **Order Apiales**

250. The **Order Dispacales** includes mostly flowering woody plants, vines, and shrubs. Honeysuckle, viburnum, arrowwood, snowberry, scabious, twinflower, and weigela are all members of this order.

251. Plants in this order have leaves that usually have oppositely arranged leaves. Many species have serrated edges on their leaves. Many members of the order have glands on their leaves that release essential oils.

252. Flowers of this group often have fused petals that create a trumpet shape, but it is not uncommon to find plants with partially fused flower petals that make two symmetrical halves like a pair of lips. The anthers of their flowers are **distractile**, while the pistils are **epigynous**.

253. Members of this order are represented in the diagram below.

JAPANESE HONEYSUCKLE **VIBURNUM** **SNOWBERRY** **WEIGELA**

254. There are several more orders of **Campanulids**, but many of these taxa contain mostly obscure shrubs or herbs that would not be familiar to the average reader. Many only have Latin names and no common name.

255. Let's move on to discuss a few of the most prominent orders of the **Lamiids**.

256. If you are a medieval alchemist, alternative medicine hippie crackpot, or play too much Skyrim, you will take an interest in the **Order Boraginales**. There are a lot of 'Bubble, bubble, toil and trouble' potion plants in this group.

257. If you didn't catch on, many families of plants in this group range from 'tummy ache' toxic to 'kiss your *** goodbye if you don't go to the ER right NOW' toxic. Viper's bugloss and heliotrope contain nasty **alkaloids**.

258. The Boraginales contain such herbaceous goodies as the buglosses, borage, hound's tongue, heliotrope, lungwort, and comfrey. Also included in the order are the ornamental forget-me-nots and oyster plants.

259. Most plants in this order have lance-shaped leaves. Some families have opposite leaves, others have alternate.

260. Most members of this order have fuzzy leaf **trichomes** that are calcified or impregnated with silica. A lot of them will irritate your skin and cause hives. Naturally, people want to eat them for medicinal purposes.

261. Flower stalks of many of the plants in this group coil like corkscrews. The flowers usually have five lobes on them that fuse into trumpet shapes or tube shapes. The **pistil** often has a two-headed **stigma**.

262. Certain families of plants in the Boraginales can signal bees that their flowers are depleted by changing the formula of the **anthocyanins** in the petals. When the flower changes color, the bees know not to bother.

263. The diagram below shows several members of the **Order Borginales.**

264. Order Gentianales also contains a large number of flowering herbs that make secondary compounds. In fact, it is one of the most massively large orders of plants, period, with more than 20,000 known species.

265. There are numerous Gentianales that make pharmaceutical compounds that were originally intended to deter herbivores. While some families make **alkaloids**, others make compounds called **iridoids**.

266. Some of the many drugs that come from this family are from rosy periwinkle (vincristine for leukemia treatments), psychotria (syrup of ipecac to induce barfing), and African snakeroot (rauvolfia for hypertension).

267. Don't get the idea that all these compounds are good for you. Some Gentianales make powerful toxins. The namesake strychnine trees are highly poisonous, suffocating animals by paralyzing the diaphragm muscle.

268. The curare vine makes a similar toxin used to tip poison blow darts by South American natives.

269. Rather than locking muscles in a flexed position, it interferes with **acetylcholine receptors** and stops breathing by paralyzing the unflexed diaphragm. Dead monkeys and birds then fall from their perches.

270. Oleanders produce **cardiac glycosides** and **neriosides**. The first interferes with the rhythm of heart-beats, while the latter is a neurotoxin. Eating oleander bushes is a good way to end up biting the big one.

271. Even innocent looking Carolina jasmine, which little old ladies like to grow as a decorative climbing vine makes some compounds that could kill a horse. However, at low doses, the toxin can be used to treat migraines.

272. The diagrams below show a number of toxic members of Order Gentianales.

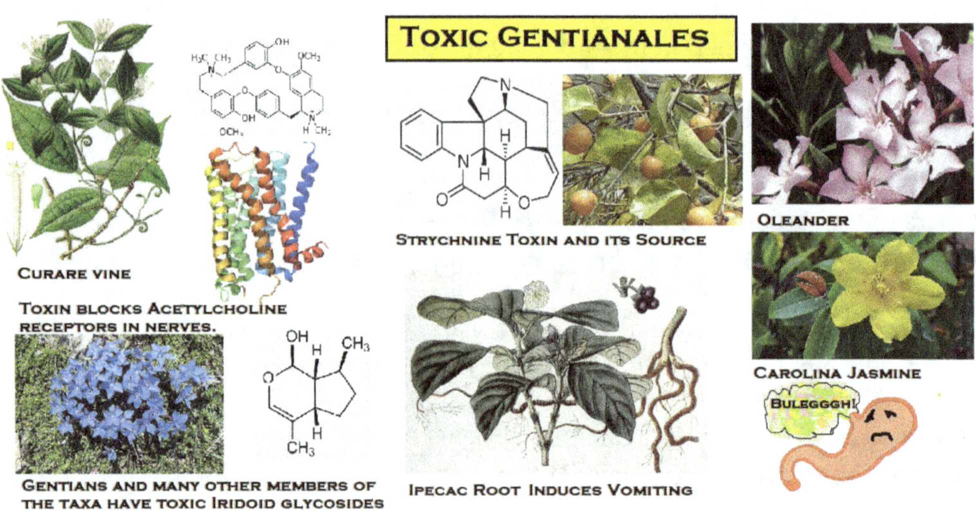

273. Members of the **Order Gentianales** typically have opposite leaves or whorled leaves with **stipules** connecting them to the main stem. Many plants in this order secrete nasty gooey **alkaloid** toxins if the leaves are crushed.

274. Gentian flowers are fairly average looking perfect flowers with predictable symmetry, but they have **didymous stamens**. They usually have compound ovaries, with the number of **carpels** depending on the family (**bicarpellary** is common), with single integuments joining dozens or hundreds of seeds to the fruit.

275. It would be remiss to not mention the indisputably most important member of the Order Gentianales. Many people would cease to function as productive human beings with coffee, which is the king of this taxa.

276. Check out coffee below in the diagram of representative members, along with other relatives that just don't matter that much, as compared to coffee. By the way, coffee is not a bean. It is a drupe.

277. The **Order Lamiales** is another massive group of plants with more than 23,000 species, represented by olives, mint, lilacs, snapdragons, ash trees, sage, sesame, basil, teakwood trees, lilacs, and lavender.

278. The plants in this order have opposite leaves. Most secrete essential oils from glands on the leaves. Unlike the previous order of Gentians, most of the Lamiales make nice-smelling fragrances that won't poison you.

279. The flowers of these plants are pretty similar to the Gentians in form. They are perfect flowers with five petals that often fuse into a tube and a pistil with two chambers in the ovary for seed development.

280. One odd quirk of this group is that they have among the smallest number of stamens in each flower of any dicot, with just four. Some have even de-evolved to smaller numbers. There are four pollen sacs in each stamen, making them **tetrathecal.**.

281. The Lamiales are also a weirdo group like the true asters, in that they don't store their carbs as starch. Instead, they make things like sugar alcohols, glycosides, and other oddball molecules.

282. In most other ways, there is such a diversity of morphological forms, flower specializations, seeds, and fruits, that it is hard to pinpoint a lot of generalities in this group, beyond the DNA evidence linking their relationships.

283. The male flower parts of Lamiales are **didynamous** and **divergent** anthers.

284. The diagram below depicts a few of the multitude of plants in the gigantic **Order Lamiales**.

285. Alright, it's about time to wrap this bad boy up. After the many pages of monocots and dicots we've covered, we haven't even scratched the surface of all the plant diversity in existence.

286. With that said, let's make the final order profiled a good one. Let's cover the order that contains the single most valuable vegetable in existence, the glorious potato. Yes, tater tots, chips, and French fries are VEGETABLES.

287. If you've learned anything from this chapter at all, by now you have also deduced that many vegetables are FRAUDS, because they are fruits that didn't have the common decency to taste sweet.

288. Beans, cucumbers, squash, tomatoes, okra, corn, eggplants....NONE of them vegetables! Arguably some of these have food value if fried or blended into a sauce that tastes nothing like their origin. However, none of these things can replace potatoes. Potatoes will never let you down.

289. So next time a self-appointed nutrition policeman decides that they are intellectually and morally superior to you because they are eating a salad and steamed vegetables when you are eating fried chicken, stop them.

290. Do not allow them to lecture you about how healthy vegetables are, because there is a good chance many of their vegetables are NOT vegetables. As long as you have potatoes on your plate, you are covered. The guide below is here to help you select true vegetables and avoid fruits in masquerade.

291. The potato is a member of the **Order Solanales**, which actually contains some pretty nasty relatives like tobacco, nightshades and Jimson weed. However, it also is home to tomatoes, tomatillos, and peppers.

292. Members of this group are known for making various **alkaloids** like solanine, nicotine, and atropine.

293. Numerous alkaloids have medical applications, in spite of their potential toxicity. For instance, atropine from the Belladonna nightshade plant is used by optometrists to dilate pupils.

294. Other alkaloids intended to detract herbivores have back-fired in some cases. For instance, capsaicin is supposed to deter us from eating peppers, but we do exactly the opposite and try to torture ourselves with ghost and scorpion peppers.

295. We all know the deal with nicotine. It causes cancer, its highly addictive, and it rightfully makes you blow chunks the first time you try tobacco, but in spite of all these hints the plant is trying to tell us, we carry right on vaping like idiots.

296. Some alkaloids like digitoxin (neurotoxin that can disrupt heartbeat) and scopalamine (central nervous toxin) are pure nastiness and do an effectively lethal job of delivering the message that foxglove and jimson weed are not to be trifled with.

297. The diagram below shows the molecular structures of several alkaloids produced by members of Solanales.

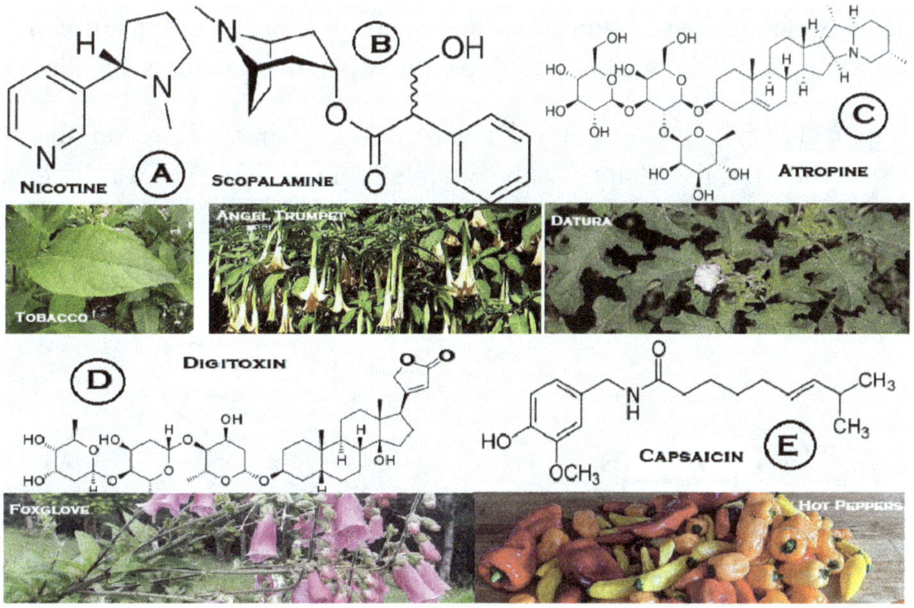

298. Most Solanales are fibrous herbs, vines, or woody shrubs. Their leaves are alternate, with many species having paired leaves of mismatched sizes that change shape as they mature. Lobed leaves and trichomes are common.

299. The flowers have **hypogynous** pistils. The stamens are connected to nectar-filled discs at their base. The ovaries are usually fused and **bilocular**, with the resulting **berry** or **capsule** fruit being full of numerous seeds from different pistil chambers. The seeds show **axillary placentation** inside the fruit.

300. Interestingly, many species of Solanales have nocturnal flowers that are bright white to reflect moonlight and draw pollinators toward them. This attracts pollinating moths and bats.

301. As you would presume, based on the potato, tuberous storage roots full of starch are common among this taxa.

302. The diagram below shows several members of this large taxa. Go get the butter and sour cream....but not for nightshades...

303. So there it is....your crash course tour through the entire phylum of flowering plants.

304. Now it's time to move on to a discussion about seeds and fruit, since this type of knowledge will further underscore an understanding for the strategies used to classify plants into orders, families, and genera.

305. It's undebatable that learning botany is very detailed and can be hard. However, that's life. Speaking of hard situations, I leave you from this chapter with a couple of unwinnable awkward scenarios that you will inevitably encounter at some point during your adulthood.

306. Inevitably, this will happen in your professional life and there are only ways to lose this scenario.

307. It is a zero sum game. Sitting there and taking it is clearly a loss. Laughing will make you look immature. Reacting angrily or with disgust will make it seem like you lack discretion and self-control.

308. The worst possible scenario is if YOU are the guilty party and everyone knows it. If you laugh, people will think you're a selfish psychopath. Your only hope is to insinuate that someone else did it and get people to believe you.

309. Let's just be real. Newborn babies are OFTEN not cute. You can just go ahead and take the loss. You are under no obligation to lie to parents about the weird phenotypic combinations that their genetics have created.

310. The odd paradox is that weird-looking kids seem to be more common if they have two ultra-attractive parents. For some reason, pro athletes and supermodels are at risk of having a child who needs to wear a bag over their head to go to kindergarten without scaring their classmates.

Chapter Five

Angiosperm Reproduction

A **) Angiosperm Seed and Fruit Development**

1.Let's move on to a discussion about the general things that happen after the ovule gets fertilized by the pollen and the seeds and fruit develop.

2. There is a different points of emphasis on development of different areas of monocot seeds versus dicot seeds, based on what will be used to nourish the developing embryo.

3. First, let's discuss the tissues that comprise an angiosperm **seed**. We will move from the outside in.

4. The **seed coat** is derived from the **integument** layers that holds the developing seed in the ovary. Eventually they feel with tough fibrous ground tissue, such as **sclereids** and form a tough dead protective layer.

5. The **seed coat** seals in moisture and protects the seed from the environment. It must be broken to allow the seed to **germinate**. In many species, this happens osmotically, when water softens the coat and swells the seed.

6. Inside of the **seed coat** is a huge amount of stored food, in the form of starches, sugar alcohols, inulin, or other carbohydrates, along with various nutritive proteins, fats, and oils.

7. Seed coats have a distinct circular mark called a **hilum**. This is the place where the **seed** was attached to the **fruit** by its integument. The part of a **fruit** that attaches to the seed is known as the **funiculus**.

8. These structures are especially obvious in any kind of legume pod and in many pepos.

9. It is easy to see all these structures in a bell pepper. The white spongy core is the integument. The strings that attach to the seeds are funiculi and the seeds themselves curve into an inside point where the hilum is found.

10. To make an analogy with animals, the **hilum** is like your belly button and the **funiculus** is like the umbilical cord. The integument is the counterpart of a mammalian placenta.

11. The **micropyle** is also visible in many seeds as a pinhole above the **hilum**. It is particularly noticeable in black-eyed peas. This is where the **pollen tube** grew into the ovule and fertilized the egg.

12. Black Eyed Peas....sigh....an abomination that cursed the world of music with autotune, musical masterpieces such as 'My Humps' and 'Boom Boom Pow' and lots of bright flashing lights and video game noises.

13. We must remember the horror, lest we ever let it happen again...'Imma be, imam be, imma be, imma be (repeat 4x), Boom-boom-boom gotta get get (repeat 4x). Somehow this became the #1 song in America in 2009....May our entire country hang its head in shame over this dark time in our history.

14. Monocots and dicots take a different approach to nourishing their developing embryos, with very different tissues responsible for the same task between the two different groups.

15. In **monocots**, the developing **endosperm** is stuffed full of starch and oils and serves as the major nutritive source. The endosperm originates from the **megaspore mother cell** just like the ova.

16. The endosperm is also a distinctly different structure from the first developing leaf, the **cotyledon**.

17. Originally, **haploid** like the rest of the **gametophyte**, it develops into a **diploid** endosperm germ line cell when a round of mitosis occurs and two of the haploid cells fuse together in the middle of the ovule.

18. When the pollination occurs, one sperm fertilizes the egg, while the other keeps going and fertilizes the endosperm germ cell, creating a **triploid endosperm**. Its sole purpose is to help the embryo develop.

19. Monocot **antipodal cells** use active and passive transport to fill the endosperm up with starches, other carbs, protein, and lipids. Like a mom packing a lunch for their kid, they make sure the embryo has plenty to eat.

20. The **endosperm** persists for a long time in monocot. You know them as corn kernels, wheat grains, oats, and rice grains. Until an **embryo** has gotten on its feet and started photosynthesizing, it feeds on the endosperm.

21. **Dicot** endosperms develop INSIDE of two **cotyledons**. Dicots develop two such seed leaves.

22. These two cotyledons, rather than being used for their conventional purpose as leaves, are stuffed with a junk food stash from the endosperm. The endosperms become indistinguishable as the cotyledons develop.

23. So to summarize, **monocot seeds** have only one food storage kernel, due to the presence of the **endosperm**. Additionally, they have only one **cotyledon** (seed leaf) that is SEPARATE from the endosperm.

24. In **dicots**, the endosperm is absorbed by **two cotyledons**. Dicots have two kernels in their seeds (the swollen seed leaf cotyledons). Unlike monocots, these won't serve a primary purpose as the first photosynthetic leaves.

25. The dicot embryo consumes the food reserves in the cotyledons as it develops. By the time the food reserves have been depleted, the seedling has grown enough leaves to produce its own food by photosynthesis.

26. Plant **embryos** have three distinct organs that will further differentiate as they grow. The **epicotyl** will become the shoot that grows leaves, the **hypocotyl** will become the stem, and the **radicle** will mature into root tissues.

27. In **monocots**, the epicotyl and hypocotyl usually fuse into one structure called the **coleoptile** that emerges as the first spike-like sprout from the soil. Secondary leaves then grow from the shoot **meristem**.

28. The diagrams below depict the tissues of monocot and dicot seeds.

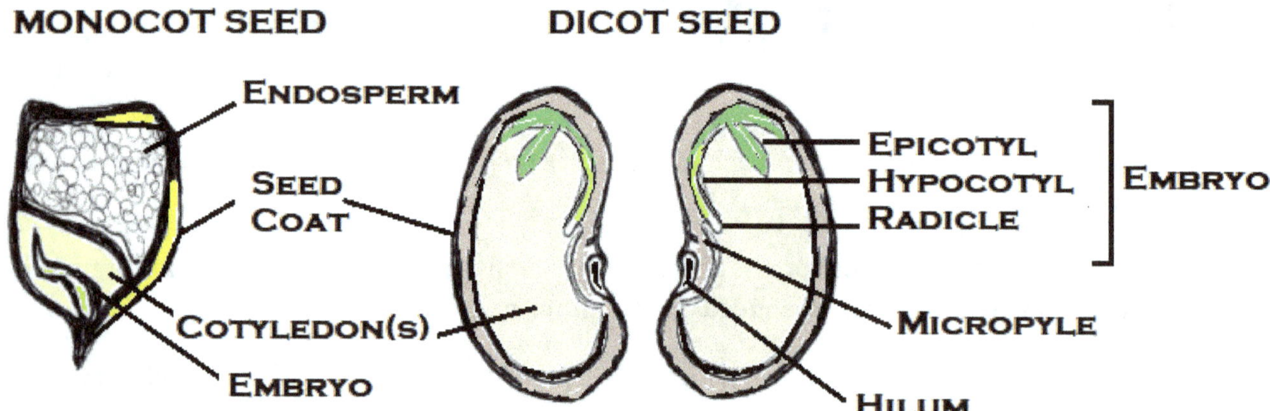

29. Without **fruit**, the seeds of angiosperms would have a much more difficult time surviving to development. After all, seeds are often full of the very same food molecules that animals also seek out.

30. Fruit gives a lot of plants a fighting chance to distribute their offspring somewhere else where they won't be direct competitors and, if the seeds are consumed, to grow in a pile of fertilizer when the animal takes a dook.

31. Many **fleshy fruits** are a bribe for transportation. Animals get a sugary meal, while the mother plant sends her children to be carried far away by a stranger.

32. **Dry fruits** don't use the same tactics. Many dry fruits serve the purpose of protecting the seed until it gets a chance to develop. For example, coconuts may float hundreds of miles before turning up on a new island.

33. Oaks and hickories allow squirrels to gather their nuts, with the strategy being that they will forget where they buried many of them. When this happens, many embryos find an ideal place to grow.

34. Some dry fruits are **dihescent schizocarps**, splitting down the middle to release seeds as they mature.

35. Dihescent fruits may rely on wind to blow small, light seeds into place, such as dandelions or they may be shaped to catch the wind like a sale, such as the samaras of maple trees.

36. Some **schizocarps** develop stickers or burrs that catch animal fur and take the seeds on a ride to a new location. This tactic is employed by beggar's lice and sandspur grasses.

37. The fruit, itself, attaches to the **hilum** of the **seed** by a part of the **integument** called the **funiculus**. Depending on the order of plants, there may be a single attachment or multiple attachments to each seed.

38. Moving from the inside of the fruit out, the **endocarp** is a fibrous layer on the interior of the fruit that the **funiculus** arises from. It is the remnants of the integument inside the ovary.

39. Obvious examples of **endocarps** include the white core of a bell pepper, the outer wall of a peach pit, the fibrous core in the middle of an orange, and the stringy goop inside of a pumpkin that holds the seeds into the fruit.

40. Outside of the endocarp is the **mesocarp**, which may be full of sugars and starches if the fruit is fleshy, or it may develop tough fibers, if the plant isn't planning on giving away its seeds for lunch.

41. In most edible fruits, the **mesocarp** is the flesh of the fruit, such as the sections of an orange, the golden juicy layer in a peach, the spongy edible parts of a zucchini, or the gelatinous sugary part on the inside of a grape.

42. The **pericarp** is the skin, or outer layer of the fruit. Orange peels, watermelon rinds, corn husks, apple skins, and pecan shells are all examples of fruit pericarps. You want some fruit puns? Ap-PEAR-ently so!

43. The diagram of the fruits below identifies these layers of tissue.

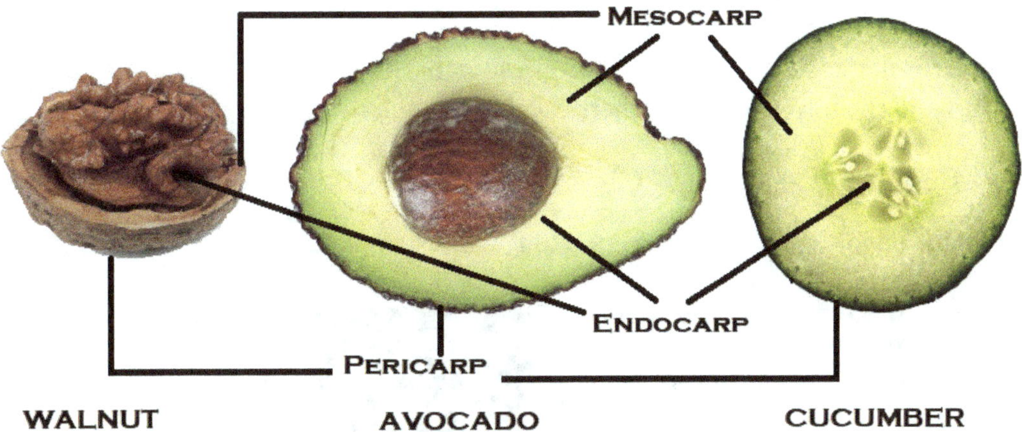

44. There are a wide variety of dry and fleshy fruits adapted for multiple different life strategies. Fruit types are not a very good predictor of classification, in most cases, because plants are so evolutionarily adaptable.

B) Types of Angiosperm Fruits

1.There are classifications of plants where reasonably close relatives produce an array of different types of fruits. Going by fruit type is as likely to mislead you in classification as it would be to help you.

2. Seed distribution to a suitable new location is the name of the game. Plants adapt whatever means is most efficient in making this happen. **Fleshy fruits** are usually designed to be bribes, while **dry fruits** usually aren't.

3. While some fleshy fruits may be toxic, this is less common than the production of palatable fruits. Many fruits are toxic to unhelpful animal species, but are not to species that distribute the seeds.

4. Toxicity doesn't help distribute seeds. For instance, every part of a tomato plant BUT the fruit contains alkaloids that can kill many herbivorous insects. No one eats tomato leaves, because they are simply nauseating.

5. Let's cover the types of **fleshy fruits** first. Descriptions and examples are given in the diagram below.

6. A fruit that has a single seed in a single simple ovary is known as a **drupe**. All stone fruits, such as peaches, plums, and cherries are drupes. Nuts like pecans would also technically be drupes if they weren't dry.

7. Many, if not most, plants have compound ovaries, resulting in the fusion of ovules and many seeds.

8. A compound fruit with a soft **pericarp** and **mesocarp** is known as a berry. Tomatoes, gooseberries, elderberries, blueberries, currants, grapes, peppers, eggplants, and bananas are all berries.

9. Larger compound fruits with leathery pericarps and more fibrous mesocarps are technically berries, but they are usually called **pepos**. Melons, cucumbers, squash, pumpkins, and gourds are all pepos.

10. **Hesperidiums** are a specialized type of berry with compartmentalized ovaries. Typically, the endocarp makes a rind that walls them off into individual chambers. Oranges, grapefruit, limes, and lemons are all hesperidiums.

11. **Aggregate fruits** are usually clusters of drupes fused together. Raspberries, blackberries, jackfruits, durians, breadfruits, and pineapples are all examples of aggregate fruits.

12. **Pomes** are what is known as a type of **accessory fruit**. In apples and pears, the stem beneath the ovary fills with sugars, swells, and grows around the true fruit in the center, which is the core of the apple.

13. Other examples of **aggregates** that really aren't fruits at all are strawberries and figs. The actual fruit of a strawberry or fig is the skin around the tiny seeds. Strawberries and figs are also swollen stems.

14. In most circumstances, fleshy fruits are meant to be eaten, so that seeds are re-distributed by animals.

15. The diagram below shows the major types of fleshy fruits, along with examples.

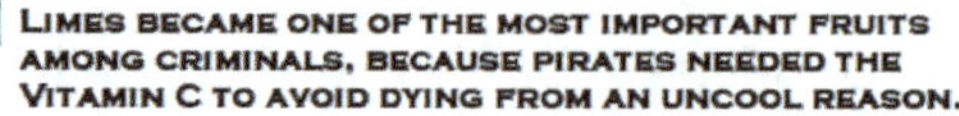

16. **Dry fruits** are typically NOT used as food bribes for animals, but as protective cases for the seeds inside.

17. Anatomically, many of the designs are the same as fleshy fruits, derived from the same tissues. For instance, nuts are just drupes that are hard as a rock. The fruit is not even close to edible.

18. Dry fruits are sub-classified as either **dehiscent** or **indehiscent**. Dehiscent fruits split down the seams to release their seeds after they dessicate, while indehiscent fruits do not.

19. Let's begin by discussing **dehiscent fruits**. There are three basic types of dehiscent fruits.

20. A simple dehiscent fruit that is derived from a single ovary is known as a **legume**. It unzips down the length of a pod-shaped fruit on both sides and dumps its beans or seeds wherever they go.

21. **Legumes** are among the only dry fruits that are sometimes designed to be eaten. For instance, the honey locust tree fills its pods with sticky caramel-like sugars that animals seek out. It's a pay to play strategy.

22. While some of the locust seeds will get eaten and die, many others will go undigested and end up in a fertilized pile of poo, while others will be scattered about by messy raccoons or possums.

23. Examples of legumes include beans, peas, lentils, locust pods, mimosa pods, tamarind pods, and mesquite pods.

24. **Siliques** are similar to legumes, but they split all the way to the stem when fully dry. Their seeds are also encased in a dry, brittle membrane called the **septum** inside the pod.

25. Cabbages, radishes, turnips, kale, broccoli, and money plants all make siliques.

26. **Follicles** are similar to legumes, but they only split open on one side when they dry out.

27. Examples of follicles include milkweed, peony, delphinium, and larkspur follicles.

28. **Capsules** are the fusion of multiple ovaries into one pod. Typically, they dump a lot of seeds. While most capsules are not edible, and some are even toxic, there are a few exceptions, such as okra.

29. Examples of plants that produce capsule fruits are sweet gum trees, lilies, poppies, cotton, and jimsonweed.

30. There are several variations of the design of capsules, based on where they split, and whether each individual ovary in the capsule splits open.

31. However, some of the examples are rather obscure and the idea remains the same in all capsules.

32. The diagram below shows examples of the different types of **dry dehiscent fruits.**

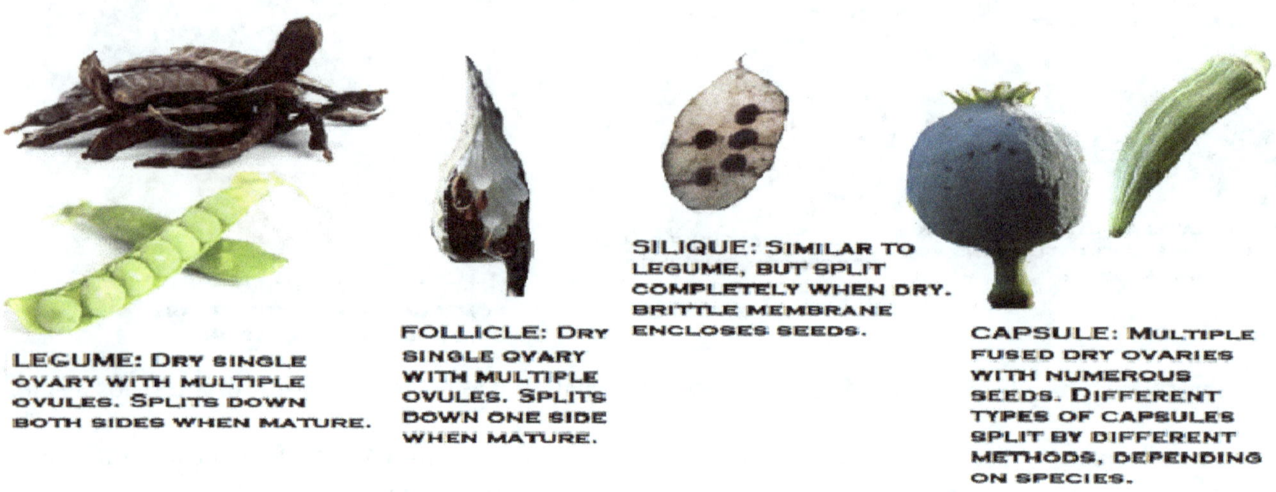

33. **Dry indehiscent fruits** do NOT split open. Either the embryo escapes from the hilum, the fruit swells and splits with the seed coat to provide an exit, or animals free the nut from the fruit or digest it away.

34. **Achenes** are simple dry fruits, with a single ovule inside of a single ovary. The endocarp and seed coat are unfused, so the seed coat and shell remain separate.

35. Examples of achenes are the seeds of sunflowers, marigolds, zinnias, buttercups, and a number of other wildflowers. The 'seeds' of a strawberry are also achenes, since the fruit is actually the covering of the seed.

36. Dandelions are a modified achene with a split pericarp designed to catch the wind and fly away.

37. A **caryopsis** is similar to an achene, but the endocarp of the fruit is fused onto the seed coat. The outer husk of such a seed is the fruit. Corn, wheat, rice, barley, oats, rye, and grasses all create caryopsis fruits.

38. A **samara** is a dry fruit that has a pericarp designed to take the shape of a wing, allowing the seed to re-distribute itself when the wind blows. Maple, ash, elm, and tulip poplar trees all produce samaras.

39. **Nuts** are the dry fruit equivalent of a drupe. The wall of the ovary is not fused to the seed. As compared to achenes, they are much larger and harder, but anatomically similar.

40. Nut-producing trees usually rely on the forgetfulness of foraging animals to plant their seeds, with the understanding that there will be collateral damage, as many of their offspring will be consumed.

41. Some nuts rely on other means of germination. For instance, coconuts rely on swelling and drying to crack through the husk, so that the embryo can emerge from the nut.

42. Brazil nut fruits are so hard that only the razor-sharp incisors of agouti pacas can open them. Like squirrels, these rodents end up planting trees, because they forget where the put some of their stashes.

43. Nuts are a common type of dry fruit. Everything from pecans, walnuts, coconuts, and mint produce them.

44. The diagram below shows representations of the different **dry indehiscent fruits**.

ACHENE: Dry single ovary and ovule. Fruit endocarp does not fuse with seed coat.

CARYPOSIS: Dry Single Ovary And Ovule. Fruit endocarp is fused to seed.

SAMARA: Dry Single ovary fused to seed. Pericarp forms a wing for wind distribution.

NUT: Dry fruit equivalent of a drupe. Ovary Wall thick and fibrous. Single large seed inside.

C) Asexual Reproduction

1. As an added bonus to all the wonderful ways flowering plants can sexually reproduce, a large number of angiosperms are also capable of asexual reproduction via **budding** or **fragmentation.**

2. Many groups of plants can easily reproduce asexually, some to the point that they rely more on **vegetative reproduction** than they do on sexual reproduction. Many groups are also incapable of asexual reproduction.

3. Asexual reproduction appears to be something that has evolved and de-evolved many different times in the plant kingdom. There is only a rough relationships between classification and ability to bud or fragment.

4. For instance, apple trees will not grow from cuttings, but pyracanthas, which are relatively closely related have no problem growing new roots. This sort of pattern is seen again and again across the plant kingdom.

5. Some entire families, such as the cacti, are almost all capable of asexual reproduction, while other entire families like the oaks, can't grow roots from cuttings and can only send out clones from roots.

6. Remember that all **asexual reproduction** comes from the creation of **clones** by **mitosis**. Therefore, root pups or cuttings produce offspring that are genetically identical from the parent plant they were derived from.

7. **Clones**, however, function independently of the parent, if broken away. With that said, some clones remain rooted to the parent root and form entire thickets of clones that can span for miles, like bamboo.

8. Cloning is a quick and easy method for plants to increase their numbers quickly and fend off competing species.

9. The diagram below lists the common means plants use to reproduce asexually and gives pictures

CORM: Swollen underground root full of starch. Can generate new clones from meristem areas.

RHIZOME: Underground horizontal stems that generates clones from meristem areas and continues to make new stem nodes.

TUBER: Underground stem that can generate clones from cuttings and meristems. Unable to spread horizontally.

BULB: Fleshy underground leaf cluster that can sprout clones from lateral meristems.

PLANTLETS: Clones of parent plants that sprout from meristems on the leaf edges. They drop off and become new independent clones.

RUNNERS: Above-ground stems that sprout from lateral meristems. Can grow roots and be fragmented to form independent new plants.

PLANTS & PLACES MEGAQUIZ: BELOVED PLANTS OF AMERICA 

ANSWER THE COMPENDIUM OF QUESTIONS ABOUT PLANTS IN AMERICA   ANSWERS ON PAGE 124

IDENTIFY THE FOLLOWING STATES BY THEIR OFFICIAL STATE TREE OR FLOWER.

1) Palmetto   2) Magnolia   3) Live Oak   4) Bluebonnet   5) Sycamore   6) Orange blossom
7) Hibiscus   8) Redwood   9) Pinyon Pine   10) Buckeye   11) Saguaro   12) Sugar Maple

IDENTIFY THE FOLLOWING COLLEGE AND PROFESSIONAL SPORTS TEAMS BY THEIR PLANT-THEMED MASCOTS

1) Buckeye   2) Cardinal (Redwood)   3) Cornhuskers   4) Shockers (Bundle of wheat)
5) An Orange   6) Maple Leafs   7) Sycamores   8) Fighting Okra
9) Cayenne   10) Fighting Pickles   11) Fleur-De-Lis Logo (Name is not a plant though)

IDENTIFY THE STATE WHERE YOU WOULD FIND THESE CITIES WITH BOTANICALLY-THEMED NAMES.

1) Oakland   2) Coconut Grove   3) Appleton   4) Redwood City   5) Cedar Rapids   6) Tupelo
7) Aspen   8) Palm Springs   9) Oak Ridge   10) Mesquite   11) Thousand Oaks

IDENTIFY THE STATE THAT GROWS THE MOST OF EACH OF THESE CROPS.

1) Corn   2) Wheat   3) Pecans   4) Peanuts   5) Oranges   6) Grapefruit   7) Lemons   8) Limes
9) Blueberries   10) Potatoes   11) Watermelons   12) Strawberries   13) Cotton   14) Apples
15) Okra   16) Carrots   17) Pumpkins   18) Sugar Cane   19) Tobacco   20) Bananas

IDENTIFY THE U.S. TREE OR CROP THAT HAS BEEN HIT HARD (PAST OR PRESENT) BY THESE PESTS.

1) Boll Weevil   2) Woolly Adelgid   3) Dutch Elm Disease   4) Cryphronectria Blight
5) Emerald Ash Borer   6) Rootworm   7) Asian Psyllid   8) Mediterranean Fruit Fly

 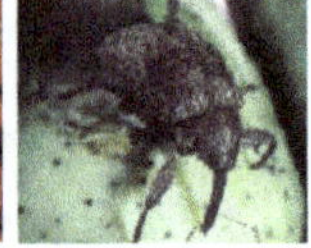     

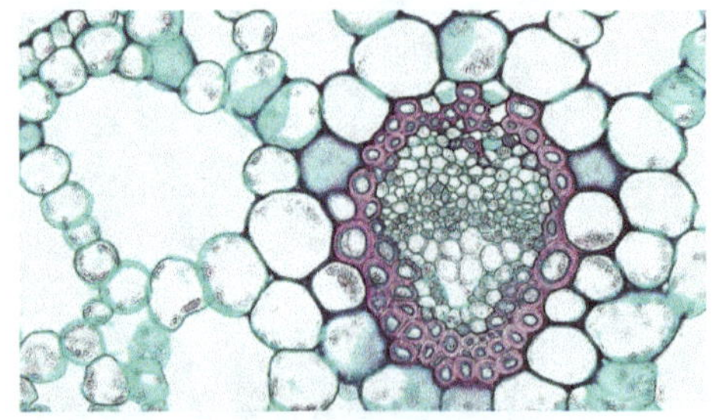

Chapter Six

Plant Tissues and Organs

S ECTION 1: Plant Tissues

A) Plant Body Plans and Life Strategies

1.**Plants**, like animals, have adapted and evolved according to their environments. Plants, like animals, are a product of their ecological niche and selective pressures.

2. In some ways, plants show much more genetic flexibility and adaptability than animals, such as their ability to asexually reproduce easily and **polyploid** genomes that produce more genotypes.

3. The combination of genetic flexibility and varied habitats and environmental pressures has created an extraordinarily diverse kingdom, with members as different as cacti and redwood trees.

4. All plants, no matter how simple or how complex (with the exception of bryophytes) have the same basic tissue types and **vegetative** organs (non-sexual). Animals show much more variability in this regard.

5. **Reproductive strategies** reflect similar lines of reasoning in both the plant kingdom and animal kingdom.

6. Some plants are closer to being **K-strategists** with indeterminate lifespans and large sizes. Many of these species also invest a lot of energy into individual seeds, packing them with nutrients, in order to maximize the number of seedlings that survive.

7. **R-strategists** grow where the growing season is short and competition is high. For instance, wildflowers on the Alaskan tundra are **R-strategists**. The tend to make thousands of small seeds and hope someone makes it.

8. **R-strategists** are commonly herbaceous short-lived **annuals** that produce copious numbers of seeds and die.

9. Annuals are frequently found in flower gardens. Some examples are pansies, marigolds, and asters.

10. **Biennials**, such as members of the carrot family, are similar to annuals, but live for two seasons. In year one, the plant stores starch in the root. In year two, all of these reserves are devoted to seed production.

11. Annuals and biennials are strictly **herbaceous**, since they don't have time to develop wood. This also minimizes the amount of resources that the plant invests in tissue growth, allowing for greater seed volume.

12. **Perennials** make up the remainder of species in the plant kingdom. They are closer to being **K-strategists**, but it could be argued that no plant is a true K-strategists, because they still don't invest parental care like animals.

13. Generally speaking, all plants also produce many more reproductive gametes from many more ovaries.

14. **Perennials** have an indefinite lifespan and don't devote as much energy to reproduction. With this said, an oak tree can still drop thousands of acorns in any given year.

15. While some perennials, like daffodils, may go dormant and come back from the roots the following year, the majority of perennials are there to stay for an indefinite length of time.

16. **Deciduous trees** also shut down most of their photosynthesis for the winter, living off of stored sugars, as photosynthesis without the requisite ingredients wastes ATP.

17. Therefore, **deciduous trees** are, generally speaking, more evolved than the more ancient **evergreen** plant species that remain green year-round.

18. However, it is important to note that **deciduousness** would not have ever evolved in the tropics, since it would serve no purpose. You have to look at other features too.

19. As such, many species legume trees are evergreen, in spite of being very evolved, since they come from tropical areas.   Losing leaves there would be a one way ticket to starvation for the tree.

20. **Perennial plants** also have evolved **secondary growth** and develop true **wood**.

21. Any tree is ALWAYS a perennial, no matter what, as it takes a minimum of 3 years to develop a woody stem.

22. In spite of the wide range of lifespans and adaptations, you will see that three basic tissue types and three basic organ types allow for the entire diversity of forms and species found throughout the plant kingdom.

23. The diagram below shows a few of the thousands of evolutionary adaptations made by flowering plants.

EXAMPLES OF LIFE STRATEGY ADAPTATIONS IN ANGIOSPERM PLANTS

POLYPLOIDY   **ANNUAL LIFE CYCLES**   **C4 & CAM METABOLISM** **DECIDUOUS LEAVES**

24. I'm now distracted. I am currently writing in front of the TV, while my wife is watching, so I'm going to rant about TV shows. She just flipped past two hipsters who could feel the 'energy emanating' from the ghost of a plumber who drowned fixing the cistern in some abandoned house where they seem to be trespassing.

25. Seriously, why are they not more afraid of the police than a ghost with a plunger and an exposed butt crack?

26. Why is it that ghosts only seem to 'haunt' mansions and other places with architectural significance or beauty? When was the last time you heard of a haunted double-wide or poltergeists in an abandoned Blockbuster Video store?

27. I'm not even going to get started on 'monster hunting' shows that amazingly NEVER find anything, yet keep viewers coming back every week hoping to catch a fleeting glimpse of a wampus cat. Alright, back to your regularly scheduled programming.

B) Ground Tissues

1. **Ground tissue**, by mass and by volume, makes up the largest portion of plant tissues in **herbaceous plants**, so we will start there on our discussion of the three different basic types of plant tissues.

2. **Ground tissue** is like the Swiss Army knife of plant tissues. It has a number of different functions, including storage, support, and photosynthesis. **Undifferentiated** ground tissue can become many different things.

3. Thanks to sets of genes that can be flipped on and off in different patterns in response to environmental cues and plant hormones, cells of ground tissue can become many different types of cells as they grow and develop.

4. The location of ground tissue within the plant lends clues to its particular function.

5. The picture below shows the primary purpose of ground tissues in the three plant organs.

6. Ground tissue can be sub-divided into three basic categories of cells known as **parenchyma, collenchyma,** and **sclerenchyma.** The largest distinction among the three is in the number of **cell walls** in each.

7. ALL plant tissues have **primary cell walls.** This is laid down by the migration of vacuoles full of cellulose to the center of plant cells when they undergo mitosis. A **cell plate** forms new cell walls around each daughter cell.

8. Though made of cellulose, primary cell walls are still fairly thin. The new cell remains flexible and pliable.

9. As the cell matures and ages, **secondary cell walls** are added on top of the original wall. **Pectin** and **lignin** help glue these additional layers of **cellulose** on top of the previous cell wall layers.

10. All of these molecules are complex carbohydrate **polysaccharides**. A diagram of their molecular structures is shown below, along with references to where they appear in the plant.

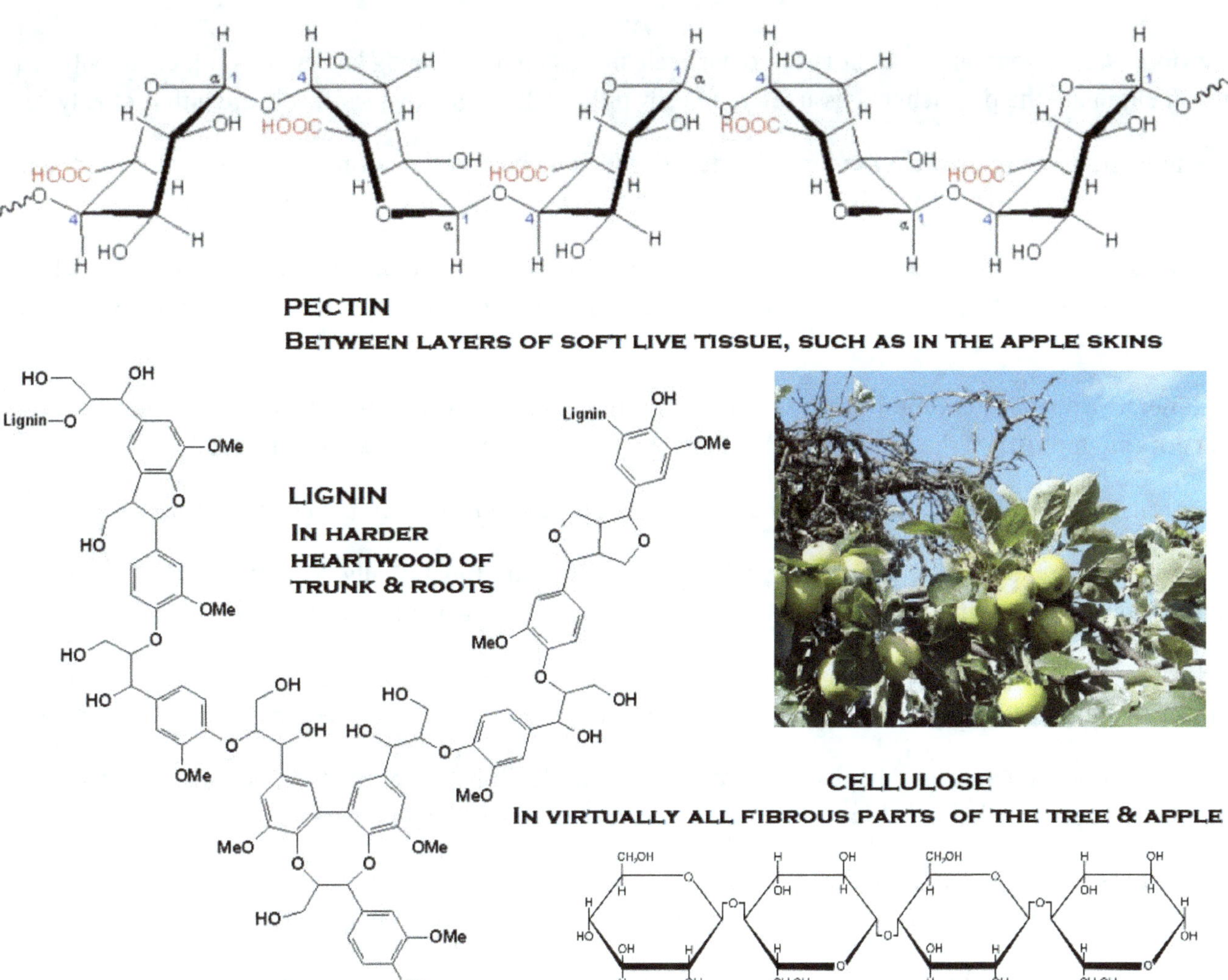

PECTIN
BETWEEN LAYERS OF SOFT LIVE TISSUE, SUCH AS IN THE APPLE SKINS

LIGNIN
IN HARDER HEARTWOOD OF TRUNK & ROOTS

CELLULOSE
IN VIRTUALLY ALL FIBROUS PARTS OF THE TREE & APPLE

11. Eventually older plant cells become inflexible and lose their pliability as they add more **secondary cell walls**.

12. Add enough of them, and it will choke the cell off and kill it. Some tissues are purposely programmed to do this, because the cells are more useful dead than they are alive.

13. For instance, stacks of **xylem** cells hollow out in the middle to form vessels that transport water.

14. The simplest ground tissue with the fewest cell walls is **parenchyma**. Parenchyma cells must be alive to perform their proper function, regardless of which plant organ that they appear in.

15. Parenchyma cells are generally round or oblong, soft, and spongy.

16. In **leaves**, the primary purpose of the **spongy parenchyma** and **palisade parenchyma** is photosynthesis. Both types of cells are full of photosynthetic pigments to run the light reactions.

17. In succulent plants, some of the parenchyma may be modified to store water or carbohydrates.

18. In **herbaceous stems** or the immature stems of woody plants, the parenchyma tissues are used for carbohydrate storage, water storage, and to house vacuoles full of secondary compounds and toxins against herbivores.

19. **Woody stems** generally have little to no parenchyma, as the developing layers of wood, eventually crush out and eliminate the parenchyma tissue in the fleshy pith, as they age. By year three, the pith is mostly gone.

20. Among roots, the **parenchyma** forms **cortex tissue**, which is used to store starch. It is also modified to help the plant move water through its tissues to reach the vascular tissues deeper in the root.

21. **Collenchyma** tissue has several layers of cell walls and has a much more rigid texture than parenchyma. If you want a good example of collenchyma, pull the strings out of a stalk of celery, because that's what they are.

22. Generally speaking, **collenchyma** is used for structural integrity in all three plant organs. **Leaves**, **stems**, and **roots** all impregnate their tissues with collenchyma to form a tough networked matrix.

23. Fibrous plants like sisal and hemp that are used for rope and fibers are full of collenchyma.

24. **Sclerenchyma** cells have many layers of cell walls. They are functionally dead at maturity, because the many layers of cellulose prevent the cell in the middle from exchanging materials.

25. Sclerenchyma makes extremely hard fibrous growths called **sclereids**. Peach pits, walnuts, coconut shells, and palm trunks are full of sclereids. Their rock-hard texture prevents damage to seeds or stems by herbivores.

26. There are several different shapes of sclereids, each tailored for their purpose. Sclereids are star-shaped, scale-shaped, or irregular and oblong, but they are not round or swollen like parenchyma.

27. **Sclereids** are used for structural supports in leaves, preventing grazing by herbivores, and to further support vascular tissues. In stems and roots, they form highly rigid fibers that prevent grazers from snacking on them.

28. Many different **dry fruits** are loaded with sclereids to protect the cotyledons and developing embryo.

FRUITY PUNS

WHAT DID THE FRUIT PIRATE KEEP SAILING THE WORLD TRYING TO FIND?
BERRIED TREASURE
WHY DID EVERYONE IN THE ORCHARD STOP GOING TO CHURCH?
THE PASTOR WOULDN'T PRACTICE WHAT HE PEACHED
WHAT HAPPENED TO THE FRUIT PRINCESS AND THE PRODUCE PRINCE?
THEY GOT MARRIED AND LIVED APPLEY EVER AFTER
WHY DID THE MUSCADINES GET THEMSELVES KICKED OUT OF THE VINEYARD?
THEY WERE UNGRAPEFUL AND UNRAISINABLE

Tissue Type	Description	Use in Leaves	Use in Stems	Use in Roots
Parenchyma (Left below)	Generally round and spongy. One or a few cell walls. Pliable. Useful for a lot of different purposes.	Spongy and palisade parenchyma are full of chlorophyll for photosynthesis. In succulent plants, may store water or sugars.	In fleshy stems, like rhubarb or potatoes, they store starch, sugar, and water in the cortex and the pith. Secrete toxins against insects.	Store starch in cortex. Used to help the movement of water through the roots
Collenchyma (Center below)	Several cell walls. Cells appear boxy and have thickened corners. Still alive, but stiffened and a lot less springy. Supportive.	Fibers for structural integrity. This is why oak leaves aren't exactly delectable and tender.	Fibers for support and structural integrity. Holds up 'woody' herbaceous plants like corn and sunflowers. The 'strings' in celery are collenchyma. Yuck.	Fibers for structural integrity. This is why most roots aren't all that wonderful to eat. These give carrots their 'snap' and strength.
Sclerenchyma (Right below)	Many cell walls thick. Come in two different shapes. Cells are dead. Sclerieds are misshapen and flattened and REALLY hard. These are found in seed coats like peach pits and coconuts. Fibers are elongated and rope-like.	Support for leaf veins and leaves that have to be really tough to avoid being eaten.	Fibers for structural support. These are what is present in the 'wood' of palm trees, even though it isn't really wood at all. Present in the bark of woody plants.	Fibers for structural support.

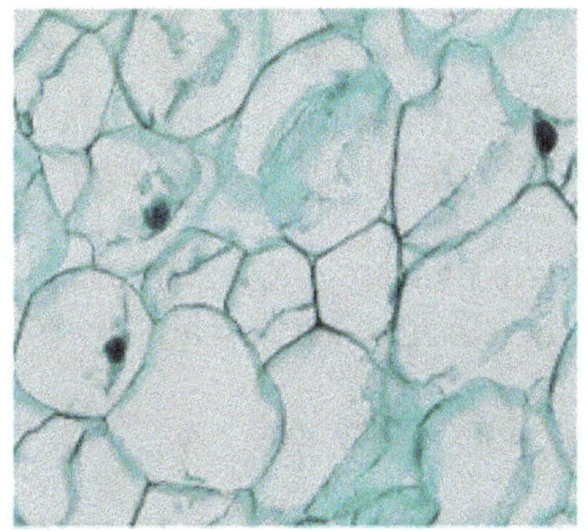

PARENCHYMA

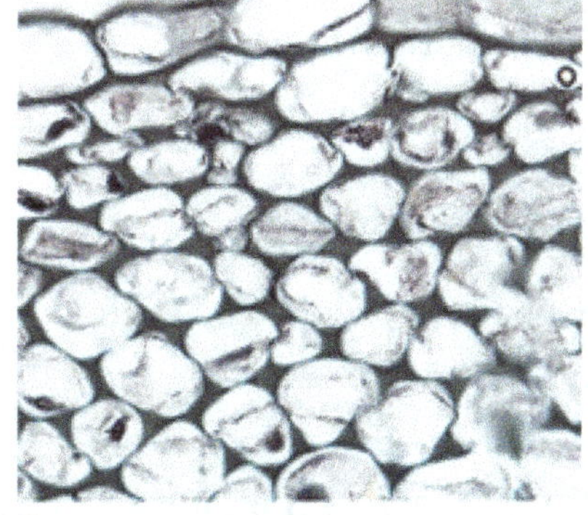

COLLENCHYMA

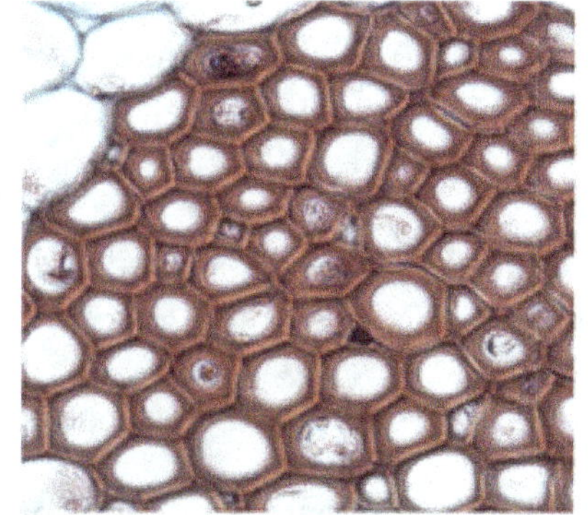

SCLERENCHYMA

29. The table above summarizes the three basic **ground tissue** types.

C) Vascular Tissues

1.**Vascular tissue** transports materials throughout the tissues of a plant. Its presence is the largest difference between a plant and a green algae, bryophytes withstanding.

2. There are two basic sets of transportation pipes running through a plant.

3. **Xylem** serves as the water pipes, taking water from the roots up to the leaves. It has many cell walls and is functionally dead at maturity. It forms millions of hollow vessels that span from the roots to the leaves.

4. **Phloem** serves as the conduit for moving sugars from the leaves down to plant tissues that need them, and finally to storage areas in the stems and/or roots. It is alive at maturity and uses **active transport** pumps.

5. Both types of tissues are produced by a type of **meristem** tissue called the **vascular cambium.** Cells in the vascular cambium are not yet **differentiated**, so they can be fated for either type of vascular tissue.

6. While different orders of plants vary the locational geometry of their vascular tissue, as a general rule, the cels on the interior of the vascular cambium divide to become **xylem tissue.**

7. Xylem continues to accumulate layers of cell walls, until transport to the cell in the middle is cut off to a point that the cell dies. After it disintegrates, this forms a hollow space that can transport water.

8. In a way, the cell gets walled off like the alcoholic Fortunato dude in Edgar Allan Poe's Cask of Amontillado.

9. Not that Poe was exactly in his right mind when he was imbibing mind-altering substances as he wrote, but there is an obvious hole in the plot here as large as the Calvin Klein-Marty McFly problem in Back to the Future (they didn't notice their son looked exactly like their high school friend and also spoke and behaved the same way).

10. Why would Fortunato hang out with a guy who was seemingly bent on revenge after he cheated his family out of their fortune and reputation? It's about as smart as hunting alone with a person you are suing.

11. So what's the Back to the Future plot hole? How do the McFlys not remember Calvin Klein, the person most responsible for their eventual marriage, and not realize he looks EXACTLY like their son years later?

12. Anyway, with all those walls, the cell in the middle dies and shrivels up, making a unit of pipe. Since they are derived from the same ring of **meristem tissue**, xylem cells stack vertically on top of each other.

13. As previously mentioned in the plant taxonomy chapter , there are two basic types of **xylem cells**. These are **tracheids** and **vessel elements**. They have fundamentally different designs that affect their conduction ability.

14. **Tracheids** are wedge-shaped and tapered on the ends and only have holes in their sides. Water must snake its way upwards in serpentine fashion, going hole-to-hole by **capillary action** through a tracheid.

15. There is not a direct vertical connection between one tracheid and the cell directly above it.

16. Tracheids are considered more primitive than **vessel elements**. **Gymnosperms** wood is almost exclusively tracheids. Primitive **angiosperms** may retain only tracheids or have a large number, as well.

17. More evolved **angiosperms** make use of mostly **vessel elements** instead. **Vessel elements**, on the other hand, are shaped like fluted pipes and stack end-to-end, allowing water to climb vertically or side-to-side.

18. Conduction in **vessel elements** is dramatically faster than in **tracheids**. This explains why pine trees are replaced by oaks and maples in temperate forests as they advance toward a **climax community** via **ecological succession.**

19. Pines simply can't compete with the greater water conductivity ability of angiosperm trees.

20. The difference is also evident in the wood of both types of trees. Softer woods like pine are almost pure tracheids, while wood that is rich in vessel elements can be extremely hard like cherry or locust.

21. The diagrams below show comparative pictures of both types of conducting tissue.

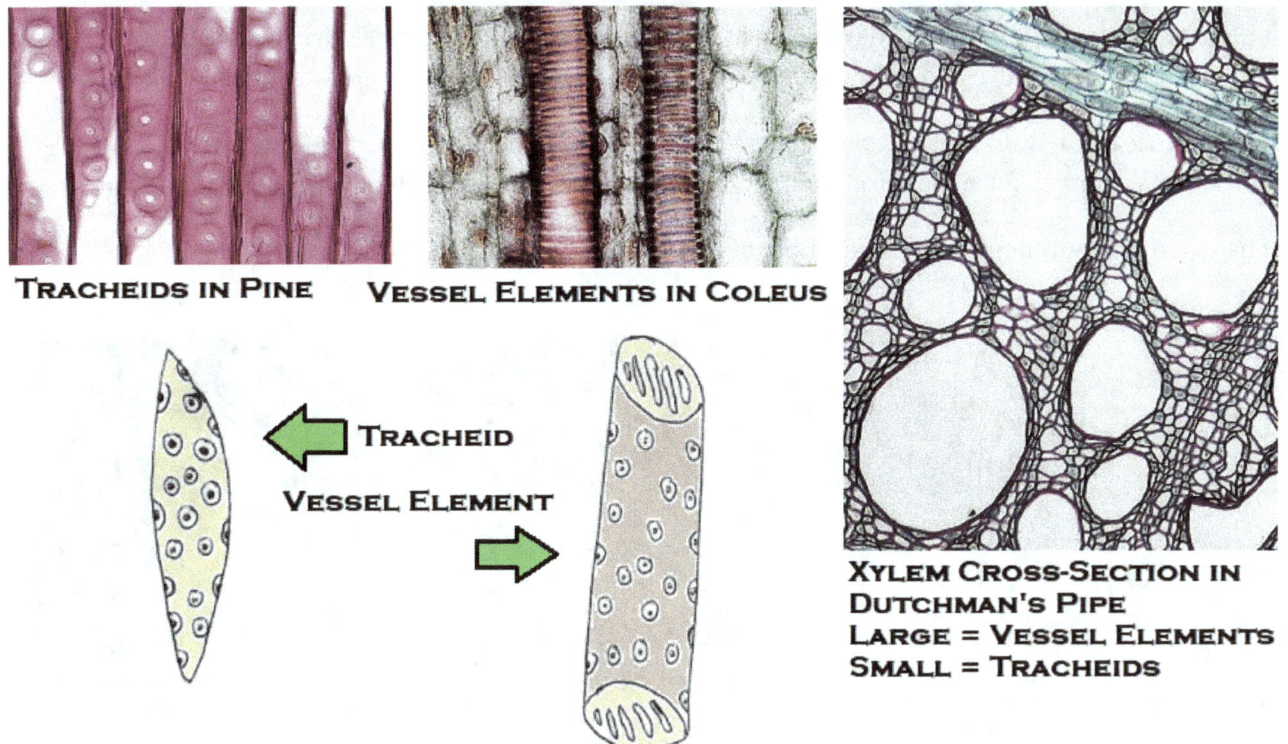

22. **Phloem** carries sugars and other photosynthetic products from the leaves to all of the tissues of the plant that need the sugars to run **cellular respiration** and do their day-to-day cell activities.

23. Whatever sugar is left over is sent to storage tissues in either the stem, root, or both, depending upon the classification of plant. These tissues are known as **cortex** (outer layers) or **pith** (interior layer).

24. The two types of cells in **phloem** are always found together in equal numbers. These cells are the hollow and helpless **sieve tube members** and their care-taking **companion cells**.

25. **Sieve tube members** are like a living doughnut. Most of the cell has been stripped away to the point that only some cytoplasm and the cell membranes are left behind.

26. Since they are also derived from the **vascular cambium** meristem layer, **sieve tubes** stack vertically on top of each other, allowing a flow of materials to drop through the open spaces inside the cell wall.

27. There are a couple of reasons that the plant needs to keep these cells alive. First, they need the **active transport membrane proteins** and **carrier proteins** on the cell membranes to function when needed.

28. Second, there are certain proteins within the cytoplasm of the sieve tubes that can still send signals to the rest of the plant to transport sugar when it is necessary.

29. **Sieve tube members** lack organelles and do not have a nucleus. This makes it necessary for each sieve tube member to have its own caretaking **companion cell** or it will quickly die.

30. **Companion cells** pump nourishment into **sieve tube members**, make proteins for them, and take over the housekeeping duties that the **sieve tube member** is not capable of doing.

31. Companion cells have lots of membrane pumps, but they are also connected to the sieve tube member with many **plasmodesmata** through each cell wall. Materials can flow freely to the helpless member.

32. The sieve tube cells, themselves, have many **plasmodesmata** at their vertical surfaces, bridging the column with other sieve tubes above and below them, allowing a pipeline of sugars to flow from cell-to-cell.

33. These vertical cell walls look like someone blasted them with a shotgun. These specialized cell walls are known as **sieve plates**. Each plate is joined to the one above or below it, glued together with pectin.

34. The diagram below depicts the sieve tube members and companion cells of the phloem.

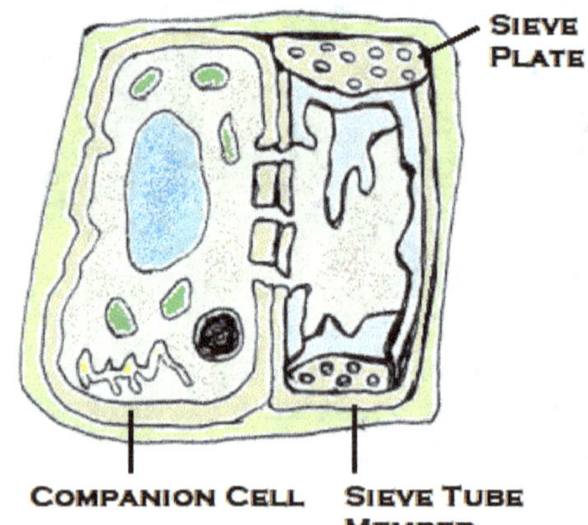

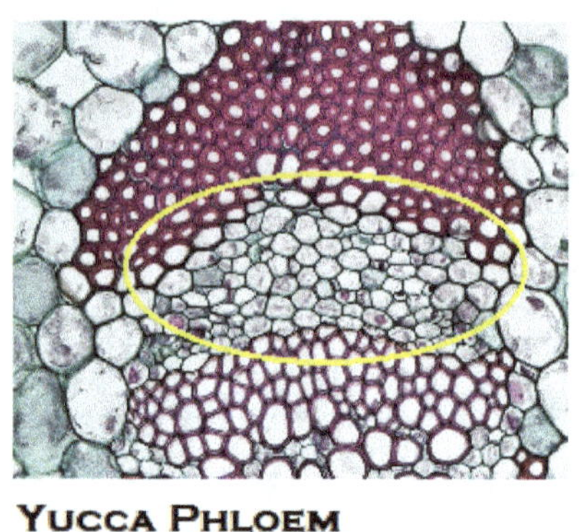

YUCCA PHLOEM

TAPPING THE PHLOEM OF A
MAPLE TREE TO GET SYRUP

D) Dermal Tissues

1.**Dermal tissue** is just what you would expect. It forms a protective skin around the plant.

2. **Dermal tissue** has similar functions on leaves, stems, and roots, but is especially important on the roots of all plants for water uptake. The cells tend to be flattened for easy diffusion.

3. **Root hairs** are extension of the root dermis that increase the water-conducting and nutrient uptake capacity by hundreds or thousands of times. They are very small, thin, and numerous, so the surface area is gigantic.

4. **Dermal tissue** is found only on the stems of herbaceous plants, but it is absent from woody stems. When trees develop wood, the **cork cambium** keeps dividing outward to make bark, which eventually splits the dermis.

5. As the tree grows, the dermis disintegrates and falls away, usually during the third season of growth.

6. On the surface of leaves, the **epidermis** protects the upper and lower surfaces of the leaf. It secretes the **waxy cuticle** to prevent water loss, and houses the **guard cells**.

7. Embedded in both layers of epidermis are **guard cells. Guard cells** are specialized **epidermal** cells that regulate the passage of gases into the leaf for use in the Calvin cycle of photosynthesis.

8. When swollen with water, **guard cells** distend and open the **stomata**, allowing gases in and out. They can control this swelling by pumping K^+ ions into their cytoplasm. This draws water into the cell osmotically. This is shown below in the diagram.

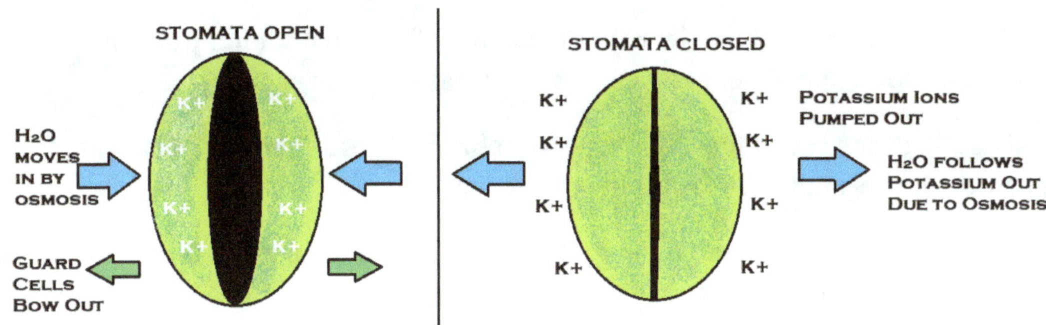

9. When plants run low on water, the **guard cells** collapse and shut over the **stomata**, preventing additional water loss from an already thirsty plant. CO_2 uptake is pointless without water to run the light reactions.

10. Some plants also modify their dermal leaf tissue by growing tiny projections called **trichomes.** Trichomes look like peach fuzz or tiny spines on the surface of the leaf.

11. Most **trichomes** are used to trap dew and get moisture. Trichomes are particularly common in plants from arid climates that border the ocean. Morning fog may be their only consistent source of moisture.

12. Examples of fuzzy plants with obvious leaf trichomes include lamb's ear, thyme, cucumbers, okra, and tomatoes.

13. However, some plants like milk thistles use the trichomes like prickly little needles to ward off being eaten. There are also glands at the base of the spines that secrete irritating compounds to drive home the message.

14. The most notorious venomous plant in the world is the Australian gympie gympie tree. It secretes a potent neurotoxin from its trichomes, which can also put the victim into anaphylactic shock.

15. Apparently the pain lasts more than a week, feels like an acid burn, and is something you'll never forget.

16. The pictures below depict several plants that have pronounced trichomes on their leave blades or petioles.

LAMB'S EAR   **CUCUMBER**   **THISTLE**   **STINGING NETTLE**

17. Dermal tissue tends to have similar functions to the leaves on herbaceous stems. Herbaceous stems also contain stomata, trichomes, and a waxy cuticle. Their structure isn't much different than a leaf petiole.

18. Dermal tissue is not present in mature woody trees because the **cork cambium** keeps dividing to make bark. This eventually splits the dermal layer and it sloughs off. This new bark layer is called the **periderm**.

19. **Periderm** is made of dead parenchyma (ground tissue), so in spite of the misnomer, it is NOT dermal tissue.

20. In the roots, **dermal tissue** is critical in all plants for the effective and efficient uptake of water and dissolved nutrient ions like potassium, nitrates, phosphates, and so on.

21. Frequently, the membranes of these dermal cells will be full of **ion channel proteins** and various types of **symports** and **antiports** that either passively or actively take up nutrient ions.

22. Root dermis is also divided into thousands of tiny **root hairs**. These dramatically increase the surface area of roots by many thousands of times, allowing the tree to maximize fluid uptake.

23. **Mycorrhizal fungi** also live on and in the root hairs. These fungi branch off of the root hairs and add exponentially more surface area for water and nutrient uptake.

24. Many species of plants cannot survive without their root mycorrhizae, because they die of malnourishment or poor water uptake over a period of time. This is particularly true for orchids and conifers.

25. The pictures below show the dermis in leaves, herbaceous stems, and roots.

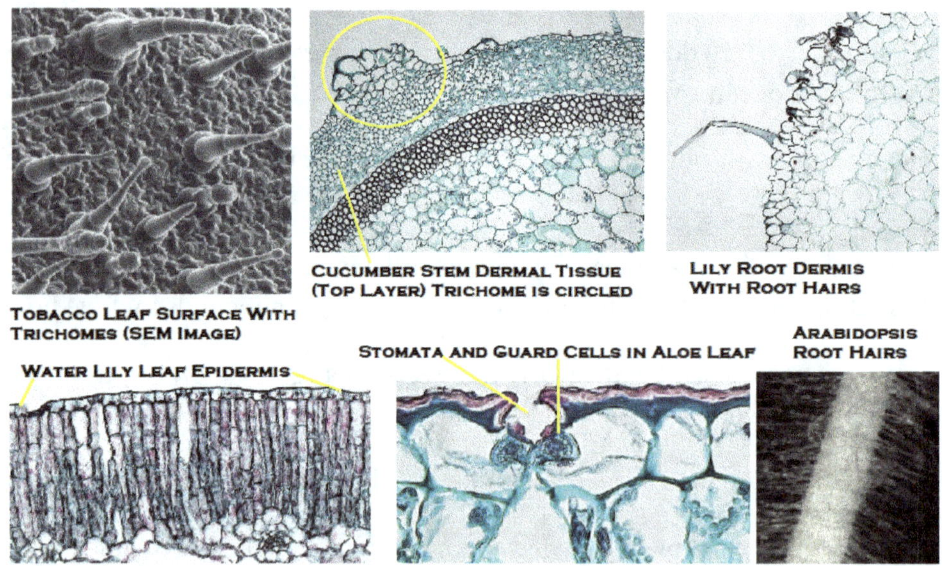

E) Meristem Tissues

1. **Meristem tissues** are **undifferentiated** cells in plants where growth starts. Therefore, **meristem tissues** can develop differentially to become ground, vascular, or dermal tissue.

2. Basically, since the cells are undifferentiated, the genes haven't switched on or off yet to tell them what their fate is going to be. **Meristem tissue** is the equivalent of stem cell tissues in animals.

3. In addition to genetic blueprints, numerous stimuli like plant hormones, neighboring cell type, light, nutrients, and many other triggers are responsible for turning these genes on and off.

4. **Apical meristems** are growth centers found at the tips of plants on the shoots and roots. They are centers for **primary growth**, meaning that they are locations of division that causes the plant to lengthen.

5. **Apical meristems** located at the end of the shoot tip are known as **terminal buds**. These growth centers divide and give rise to tissues that cause trees to grow taller.

6. Apical meristems at the **root cap** cause roots to lengthen and push deeper into the soil.

7. **Lateral meristems** are called **primary meristems**. They are found on the sides of stems. These lateral meristems are what cause plants stems or trunks to grow side branches and get bushier.

8. However, there are two other types of **secondary meristem** tissues in stems known as **cambium layers**.

9. Cambium layers occur in concentric rings on the inside of stems and roots.

10. There are two types of cambium tissue layers that occur at different depths inside the stem.

11. **Cork cambium** divides to the outside to become a type of ground tissue known as **periderm**. Periderm is the corky outer layer of bark that protects the deeper layers of the tree trunk. It is dead at functional maturity.

12. **Vascular cambium** is deeper in the trunk. It divides on both sides, generating **phloem tissues** toward the outside and producing layers of **xylem** on its inside edge to create new rings of wood each growing season.

13. All three types of meristem tissue are also found inside of roots.

14. **Lateral meristems** in roots cause may cause to grow side roots and spread out laterally in the soil.

15. In roots, both cambium layers essentially do the same thing that they do in stems. However, once these divide and produce bark or wood fiber, that root is essentially useless for nutrient uptake.

16. Roots that have **secondary growth** (cork and wood layers) are only used as support roots for large trees. Once they are impregnated with woody tissues, their ability to conduct water is minimal.

17. Smaller tender roots deeper in the soil, with intact dermis and root hairs, do the conducting for trees. They are greatly aided by mutualistic fungi growing as **mycorrhizae**.

18. The pictures below show the various types of meristem tissue.

LILAC TERMINAL BUDS APICAL MERISTEM

FLOWERING QUINCE LATERAL BUDS LATERAL MERISTEMS

BLACKBERRY CORK CAMBIUM

TERMINAL AND LATERAL BUDS IN LUPINE ROOTS

PINE VASCULAR CAMBIUM

SECTION 2: Plant Organs

A) Leaf Anatomy and Histology

1. You have probably known this since you were three years old, so this will come as no earth-shattering revelation...**Leaves**, in most cases, are the primary organs in plants where **photosynthesis** occurs

2. **Leaves** are basically nature's solar panels. The basic macro-anatomy of a leaf is shown and described below.

3. Now let's zoom in for a look at the micro-anatomy within the leaf tissues. It's all about photosynthesis.

4. The entire function of the leaf comes down to the activities inside the **choloroplasts**. Without the activities of **plastids**, there would have been no real purpose for plants to have evolved leaves.

5. Let's quickly recap what happens during the two stages of photosynthesis. First, the **light reactions** split water and use electrons and protons from hydrogen atoms to set up an energy gradient along the inner membrane.

6. Using this **electron transport chain**, the electrons and protons chase each other down a series of electron carriers, generating an electrochemical gradient. At the end of the chain, is an **ATP pump**.

7. The protons, initially pumped into the space between the inner and outer membranes of the chloroplasts, come rushing back in through an opening in the ATP pump, rejoining the electrons they have been trying to catch.

8. Thanks to the motion of these flowing protons, this allows an energy transfer to the **ATPase enzyme** that harnesses the energy like a water wheel and recharges spent **ADP** into **ATP**. This happens in **Photosystem II** of the light reactions, which actually come first, but were the second pathway to be discovered.

9. One set of light reactions makes 18 ATP that will be passed on to the **Calvin-Benson cycle.**

10. Meanwhile, leftover protons and electrons are joined to the electron carrier, **NADP+**, to form **NADPH**, which will carry the hydrogen atoms to the next stage, the **Calvin-Benson cycle**, to be used in sugar manufacturing.

11. Each set of light reactions produces 12 NADPH. This happens at the end of **Photosystem I**, which actually is second in the sequence, but was the first to be discovered.

12. These **light reactions** occur in the **thylakoid membranes** of the chloroplasts, where 'solar panels' full pigments trap photon energy and use it to excite electrons that then go pinging around down electron carriers.

13. Collectively, all the thylakoids in a chloroplast are known as the **grana.** Outside the **grana** is a soup of enzymes known as the **stroma**, where the **Calvin-Benson cycle** occurs.

14. The diagrams of the light reactions below have been included to provide a quick review for those familiar with the metabolic pathways of photosynthesis, or as a quick guide for newbies, as to how the basics of the processes work.

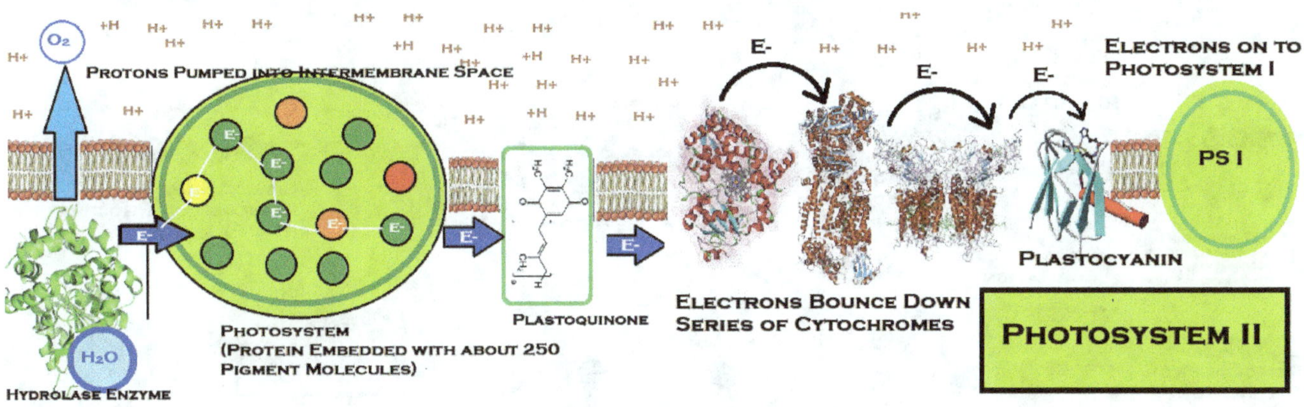

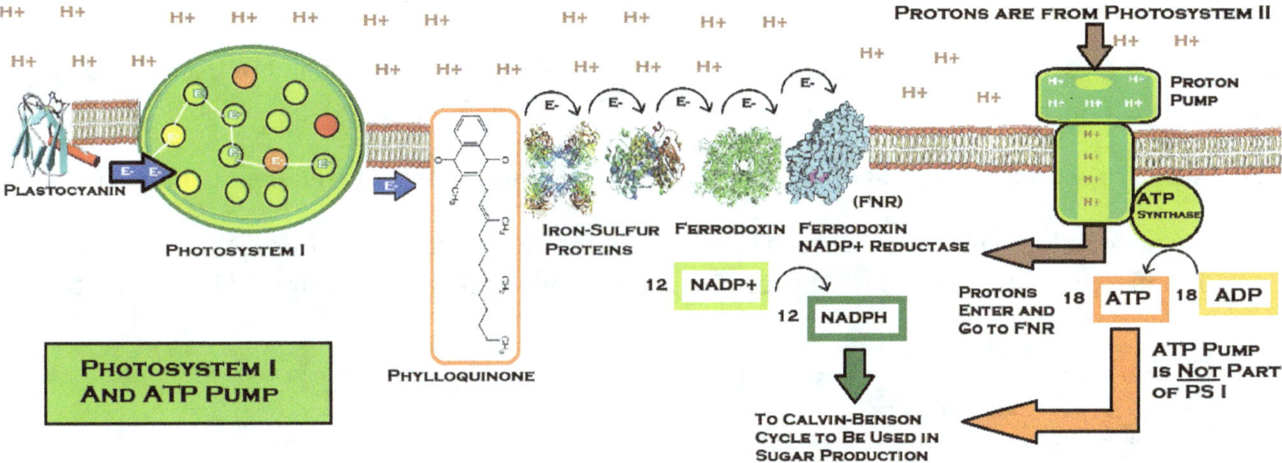

15. The Calvin-Benson cycle takes the ATP and NADPH from the light reactions and uses them in the process of manufacturing biomolecules like sugars.

16. Each turn of the Calvin-Benson cycle uses the energy from 18 ATP to form chemical bonds to make sugars.

17. The enzymes use the energy to form bonds between 6 CO_2 molecules, hydrogen from the 12 NADPH carried in from the light reactions, and recycling **Ribulose bis-phosphate** molecules from the previous Calvin-Benson cycle.

18. For a full discussion of the entire process, please refer to **Camp's Cell Biology and Biochemistry by the Numbers.** The basics of these so-called 'dark reactions' are shown in the diagrams below.

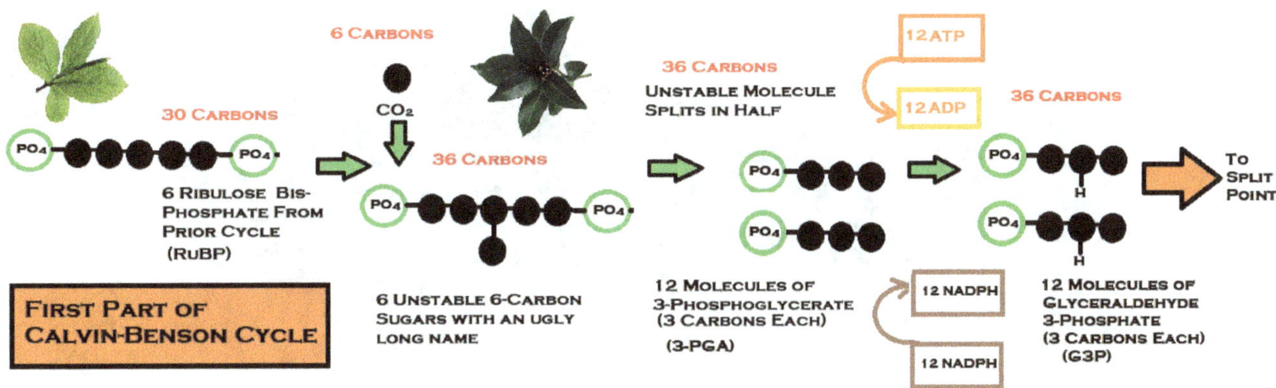

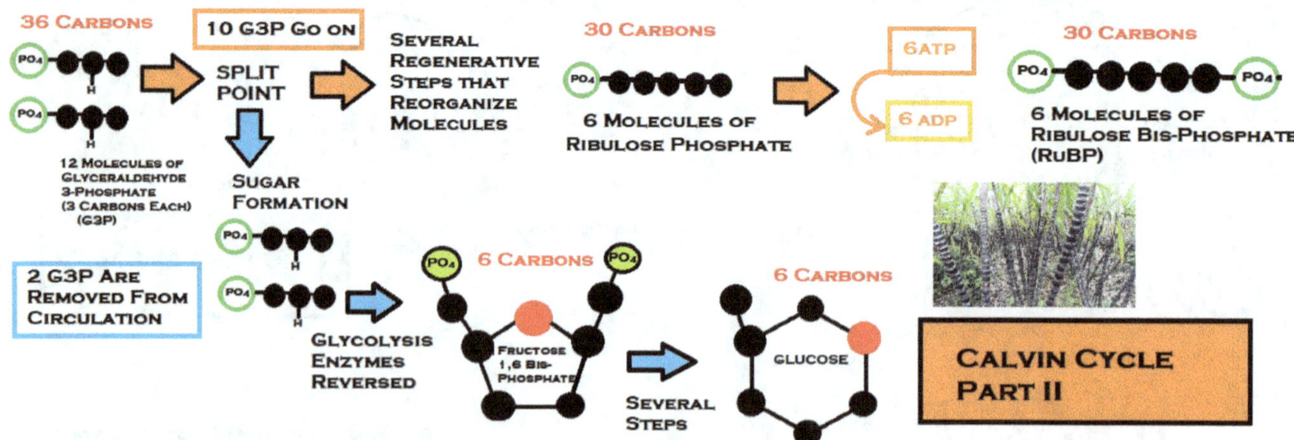

19. Within the leaves, the tissues that house chloroplasts are **spongy mesophyll** and **palisade mesophyll cells**. These occur in layers at the top and middle of the leaf, directly beneath the upper epidermis.

20. **C4 plants** also use the **bundle sheath cells** around the leaf veins to complete the steps of photosynthesis.

21. There are anatomical differences between the leaves of monocots and dicots. Many of these differences relate to the different strategies employed by regular **C3 photosynthesis** versus **C4 or CAM photosynthesis.**

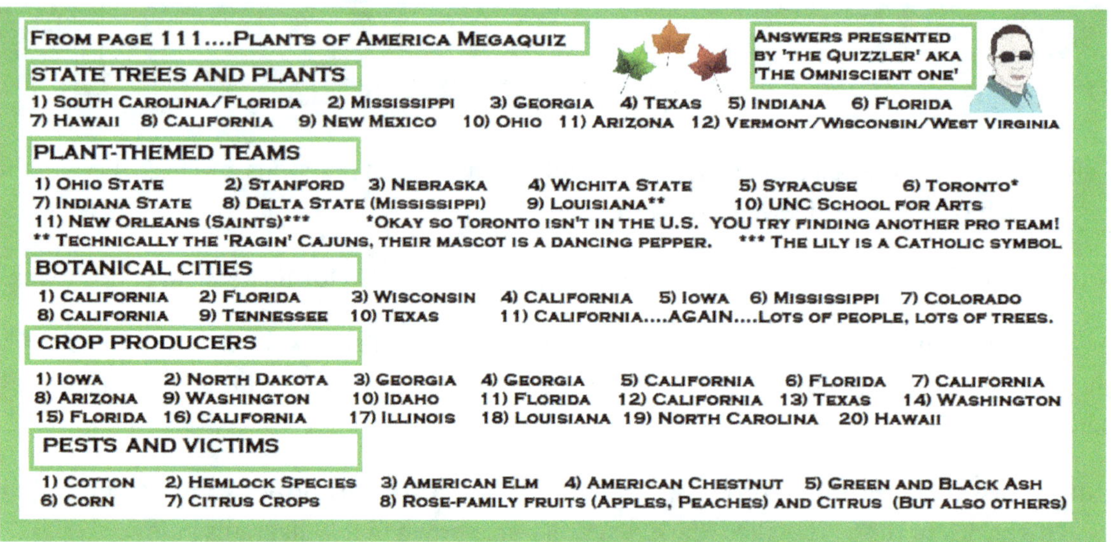

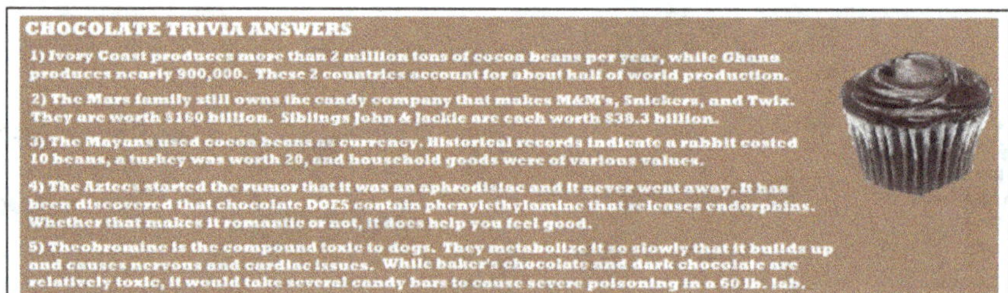

Part of Leaf	Function	Found in...	Tissue Type
Cuticle	Waxy layer secreted by epidermis. Helps plant prevent water loss.	All.	N/A
Upper epidermis	Top dermal layer on a leaf. Seals leaf from water loss and protects it.	All.	Dermal
Trichomes	Extensions of the epidermis that help the plant trap water with 'fuzz' or ward off herbivores with spines or spikes on the leaf surface.	Depends on species. Usually not present.	Dermal
Pallisade mesophyll	Picket-fence shaped cells that are found underneath the surface of the epidermis. Loaded with chloroplasts for photosynthesis.	C3 plants only	Ground
Spongy mesophyll	Less differentiated blob-shaped mesophyll cells that are full of chloroplasts for photosynthesis. Make & pump sugars into leaf veins.	C3 plants; Only layer of mesophyll in C4 and CAM.	Ground
Bundle sheath cells	Outer layer that wraps up the leaf vein. Pump and import sugars into the phloem and help deliver water to the leaf.	All. Enlarged and used for storage of PEP in C4 & CAM.	Vascular
Xylem	Transports waters from roots and stem up to the leaves for light reactions and metabolism.	All.	Vascular
Phloem	Transports sugars form leaves to other parts of the plants that need them for metabolism. Ultimately continues to stem and roots.	All.	Vascular
Lower Epidermis	Lower dermal layer of a leaf. Serves similar function to upper epidermis, but contains the stomata for gas exchange.	All.	Dermal
Guard Cells	Specialized dermal cells that form a doughnut shaped opening between them. When swollen, they open the stomata. When wilted, they close the stomata to prevent water loss.	All.	Dermal
Subsidiary Cells	These cells sit next to the guard cells and pump water into the guard cells to open the stomata. They control opening and closing.	All.	Dermal
Stomata	Openings that allow gas exchange on the underside of terrestrial plant leaves (may be on the upper surface of aquatic plant leaves).	All.	N/A

22. The table above gives a generalized description of parts and functions and discusses whether the layer is present in C3, C4 (and CAM) or both types.

23. Pictures of cross-sections of both C3 and C4 leaves follow after numbers 33 and 39 below.

24. The first one is a C3 plant, along with a picture of real C3 plant (a privet hedge leaf).

25. Now let's look at cross-sections of a typical **C3** dicot leaf with two mesophyll layers, and a **C4** monocot plant, where it is necessary to employ an enlarged layer of bundle sheath cells and only a single layer of mesophyll.

26. The first diagram hows pictures after #76 shows a leaf cross-section from a typical C3 plant, the lilac.

27. Notice that there are two distinct layers of mesophyll. The **palisade mesophyll** layer is tightly packed and each cell is full of chloroplasts. Most of the photosynthesis that occurs in the leaf happens here.

28. The **spongy mesophyll** has fewer chloroplasts, the cells are rounder, and they are loosely packed. The numerous air pockets in the spongy mesophyll let CO_2 travel throughout the leaf to be supplied to the **Calvin-Benson cycle**.

29. Spongy mesophyll also photosynthesizes, but gets less sun in the middle of the leaf and has less chloro- plasts.

30. If they were on a leaf basketball team, the palisade mesophyll would be the starter, while the spongy mesophyll would be the scrub waving the towel that gets into the game with 29 seconds left and frantically shoots airballs.

31. **Vascular bundles** transport water into the leaf via the **xylem**, while newly-formed sugars are pumped into the **phloem** vessels to be exported down to the stem and roots.

32. **Bundle sheath cells** merely wrap up the vascular bundles in C3 dicots and don't do any photosynthesis.

33. Both the **upper and lower epidermis** protect the cells inside the leaf, secrete wax, and are studded with **guard cells** that open to allow carbon dioxide into the stomata. Water and oxygen diffuse out, like it or not. The diagram below shows the structure of a C3 leaf.

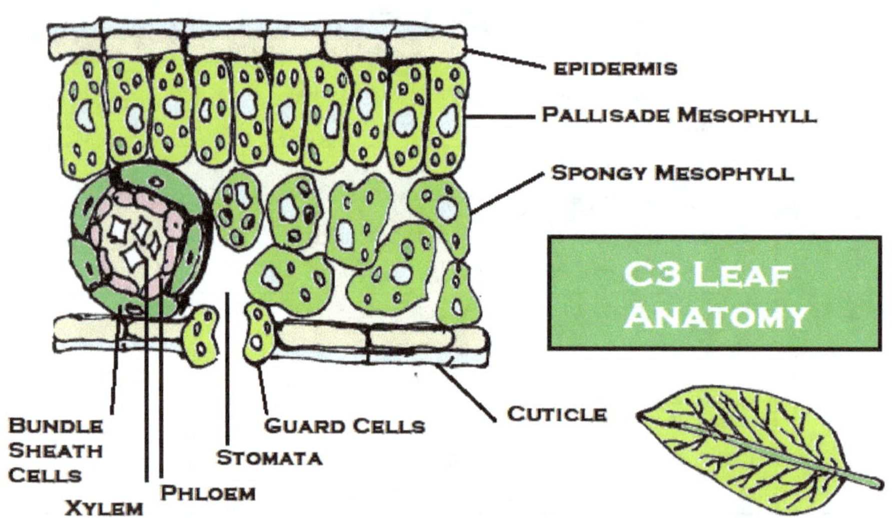

34. Now let's look at a **C4 monocot leaf**. Corn has this type of classic leaf anatomy.

35. There is only one uniform layer of **mesophyll cells** in the thinner monocot leaf, as only one is needed.

36. **Bundle sheath cells** form swollen structures that surround the xylem and phloem of the vascular system. These are full of vacuoles that store the 4-carbon **malic acid** as it is produced by the **PEP carboxylase** enzyme system.

37. Malic acid can then be enzymatically cleaved into 3-carbon **pyruvic acid** and CO_2. This allows C4 plants to have a supply of CO_2 ,without the need to open their stomata in the heat of the day and risk dehydration.

38. Corn and other C4 plants can keep producing sugars during the heat of the day when C3 plants have to stop. A schematic of the C4 biochemical pathway is shown below. Note that these plants ALSO run the C3 pathway and that C4 is a side shunt.

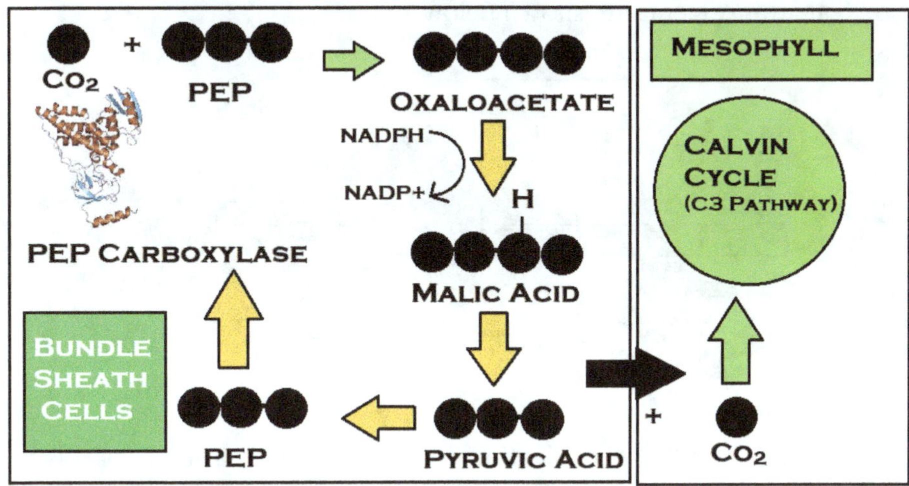

39. A cross-section of a C4 leaf (corn) is shown below.

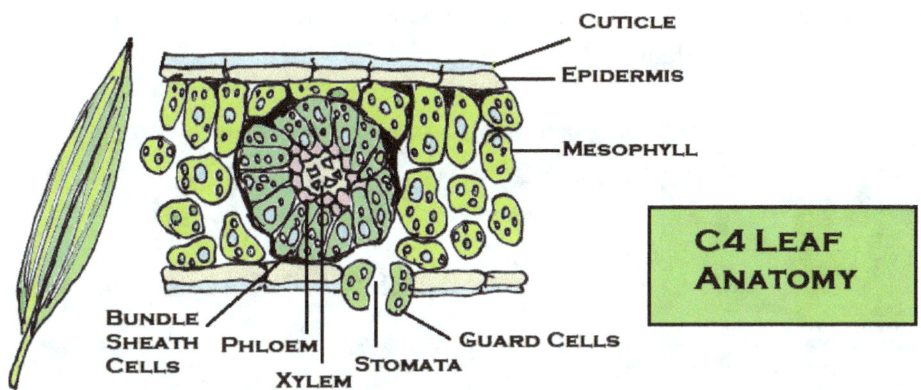

40. Now let's zoom back out of the cellular level and look at leaf macro-anatomy again.

41. Leaves, as you know from observations, come in a tremendous variety of shapes and arrangements.

42. While many plant families can be identified by leaf morphology, sometimes shapes and arrangements are evolutionary adaptations to their particular environment, rather than uniform across the family.

43. For instance, most of the members of the ginger family have broad paddle-shaped leaves to maximize sunlight collection in the forest understory. Ginger, bananas, and birds-of-paradise all have similar leaf genetics.

44. While these two plants are good instances of leaf shapes being retained within the same classification, many families of plants have highly diversified leaf types among genera, based on environmental pressures.

45. For instance, oaks have a huge diversity of shapes. Understory blackjack oaks have large fiddle-shaped leaves larger than a man's hand, while the leaves of giant live oaks are as narrow as a pencil and very small.

46. While the majority of plants in nature probably tend to have **simple leaves** (I'm not going to try to count), with one leaf per **petiole, compound leaves**, with multiple **leaflets** on a single **petiole** are also very common.

47. **Compound leaves** can then be divided up into those that are **palmately compound** (shaped like the fingers on a hand) or **pinnately compound** (shaped like a fern).

48. **Twice-compound leaves**, with multiple stalks of leaflets on one large petiole are also common. Some off the most common leaf arrangements are shown below.

49. Leaves are extremely useful in plant classification, but it is beyond the scope of this book to go into the level of detail that a professional botanist would expect.

50. Leaf form, along with the description of the edges (smooth, serrated, lobed, etc.) can be of great assistance in identifying different plant species with dichotomous keys. Some examples of leaf shapes are presented below, with representative species identified.

53. In the great majority of cases, the essential function of leaves is to perform photosynthesis.

54. However, plants also sometimes modify and re-purpose leaves for other purposes.

55. Since leaves are full of ground tissue parenchyma (the mesophyll), there is no rule written in stone that says that they HAVE to fill those cells full of chloroplasts and photosynthesize. They can be filled with other things too.

56. Many plants, particularly **succulents**, use the ground tissue space to store water and sugars.

57. **Bulbs**, like onions and garlic, are really just big underground leaf clusters full of **leucoplasts**, where the plastids are full of **starch** granules rather than **chlorophyll**.

58. The table on the next page describes several types of modified leaves, their functions, and examples of each.

Modified Leaf	Purpose	Examples	Picture
Bud scales	Small protective leaves that close around a developing leaf on a bud.	Almost all flowering plants (oaks, hickories).	
Spines	To prevent water loss (chlorophyll is in fleshy stem instead). To prevent being grazed upon by herbivores.	Cacti, euphorbias	
Tendrils	To climb up structures (vines). Petiole does not develop a blade.	Grapes, muscadines	
Succulence	To store water and/or starch	Aloe, hens and chicks	
Leaflets	Vegetative clones of parent plant made by asexual reproduction. Drop off to form new plants.	Kalanchoe	
Petals	To attract pollinators with bright colors or otherwise help pollination	Almost all flowering plants (poinsettias, roses).	
Bulbs	Underground starch and water storage for long-temr survival.	Onions, garlic	
Carnivorous	Leaves are modified into traps to collect insects for nitrogen.	Venus fly trap, pitcher plants, sundews.	

B) Stem Anatomy and Histology

1.Plant **stems** develop from the embryonic **hypocotyl** and contain areas of meristematic tissue that give rise to other branches, leaves, and new layers of vascular and ground tissue inside.

2. **Stems** have multiple functions. Some of their critical primary functions are the conduction of materials from leaves to roots, support of the plant, and lifting leaves up to the sunlight.

3. Additionally, **herbaceous stems** store starches, other nutritive compounds, and secondary compounds in the **cortex** and/or **pith**. Sometimes storage stems are modified as underground **tubers**, as in the case of potatoes.

4. Some plants have stems or trunks that store water in ground tissues with enlarged central vacuoles. Most cacti species, baobob trees, euphorbias,and traveler's trees take advantage of this adaptation.

5. In some species native to deserts, steppes, and other arid regions, plant stems can take on partial or full photosynthetic responsibilities. Again, this also happens in cacti and euphorbias.

6. Other plant stems are used for climbing toward sunlight. Pipevines, strangler figs, curare vines, smilax briers, and numerous other lianas use the structural support of neighboring tree trunks to get to the canopy.

7. The diagram below shows pictures of several different types of stems functions.

DOGWOOD TRUNK
CONDUCTION OF WATER
AND NUTRIENTS. SUPPORT
OF LEAVES AND BRANCHES.

FENNEL
STORAGE OF STARCH AND
SECONDARY COMPOUNDS

STRANGLER FIG
CLIMBING HOST TRUNKS
TO GET TO SUNLIGHT

EUPHORBIA
STORAGE OF
WATER AND
SECONDARY
COMPOUNDS

8. In spite of differences in functional purpose, there is predictable uniformity in the design of almost all stems.

9. Keep in mind that **herbaceous plants** have soft green stems with food storage tissues, while **woody plants** have lost most (or all) of the dermal and ground tissues that herbaceous plants retain.

10. All **monocots** have either **herbaceous stems** or **fibrous stems** that look superficially like wood, but they never develop true wood tissues. Only **dicots** have the meristematic layers to develop true wood fiber and trunks.

11. If bisected, monocots and dicots stems have very different arrangements of tissues within.

12. Envision a cut corn stem or a sugarcane billet for a good example of a typical monocot.

13. The stems of **monocots** always have scattered **vascular bundles** embedded in a layer of **cortex,** which is a large mass of ground tissue used to store carbohydrates. The **dermis** then surrounds the stem.

14. Monocot stems usually retain their **epidermis** throughout their life, with the exception of certain plants like palms, pandanus, and yuccas, which grow extensive fibrous tissues.

15. In spite of the fact that there are tree-like monocots, the majority of monocots have herbaceous stems that retain chlorophyll and play an active part of photosynthesis.

16. Taller monocots such as sugarcane, bamboo, palms, and yuccas use fibrous ground tissues to support the stems.

17. Such stems are loaded with **collenchyma** and in some instances, sclerenchyma, that give them a very fibrous, tough texture that is resistant to grazing herbivores and being blown over by wind.

18. Palm trees and yuccas use a combination of collenchyma and rock-hard **sclerenchyma** to form what looks like wood (but really isn't). Their trunks are very sturdy and resistant to wind damage and grazing animals.

19. These fibers form functional trunks and allow them to grow as tall as 200 feet, in the case of wax palms.

20. A look at monocot trunks, however, shows that they have no true **bark,** the fibers do not grow in **growth rings**, nor do they have **vascular cambium** or **cork cambium** meristematic layers.

21. The growth pattern of monocots almost always starts at the **apical meristem** at the top of the plant. **Lateral meristems** are not common, though a few species can branch sideways.

22. The diagram below depicts a typical monocot stem.

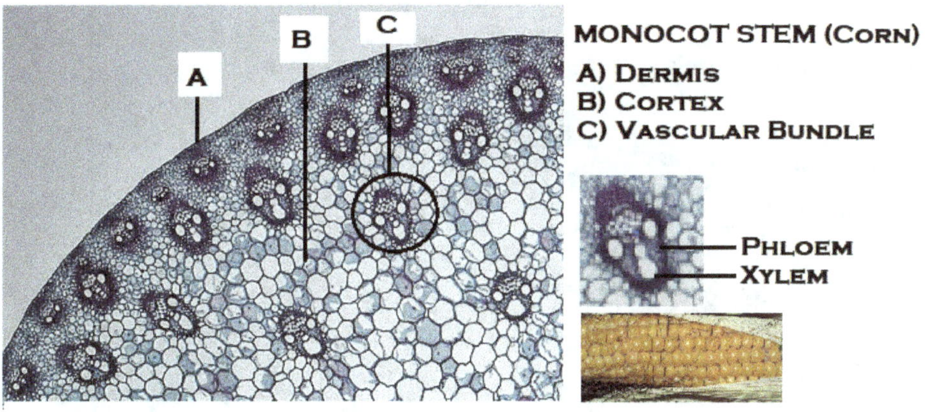

23. In corn, the **dermal tissue** is fibrous, green, and photosynthetic. The **ground tissue** in this stem is the **cortex**. It is loaded with stored carbohydrates. Finally, the **vascular bundles** are scattered throughout the cortex.

24. Now we consider the very different structure of dicot stems. While all dicot stems start with the same basic scheme, some dicots remain herbaceous, while others develop wood via **secondary growth.**

25. Let's begin by discussing the more basic herbaceous dicot stem.

26. **Herbaceous dicot stems** have a green photosynthetic **epidermis** like monocot stems. Unlike monocot stems, **lateral meristems** are relatively common, allowing side branches and nodes to form off of a main stem.

27. Many herbaceous dicots are **annuals** and **biennials** and remain **herbaceous** because their lifespan is so short that there would be no point to growing a woody trunk, since they are just going to kick off in a year or two.

28. There are also numerous **perennial dicots** that never do develop wood fibers. Most use at least some type of fibrous ground tissues to support themselves as they grow larger each season.

29. Many of these perennial species don't bother with making wood because of their environmental adaptations to harsh climates. For instance, Alaskan wildflowers bloom vigorously in summer and die to the roots in winter.

30. The pictures below depict all three types of herbaceous dicots.

31. Marigolds are **annuals** that go to seed and die in one growth season, parsley is a **biennial** that stores food in year one for seed production in year two. Carnations are **perennials** that emerge in Spring year-after-year.

32. **Dicot stem** anatomy occurs in a predictable pattern. The **vascular bundles** of dicots grow in a circular dart board pattern, with the cortex outside of this, and the **pith** in the center.

33. A typical herbaceous dicot stem is pictured in the following diagram.

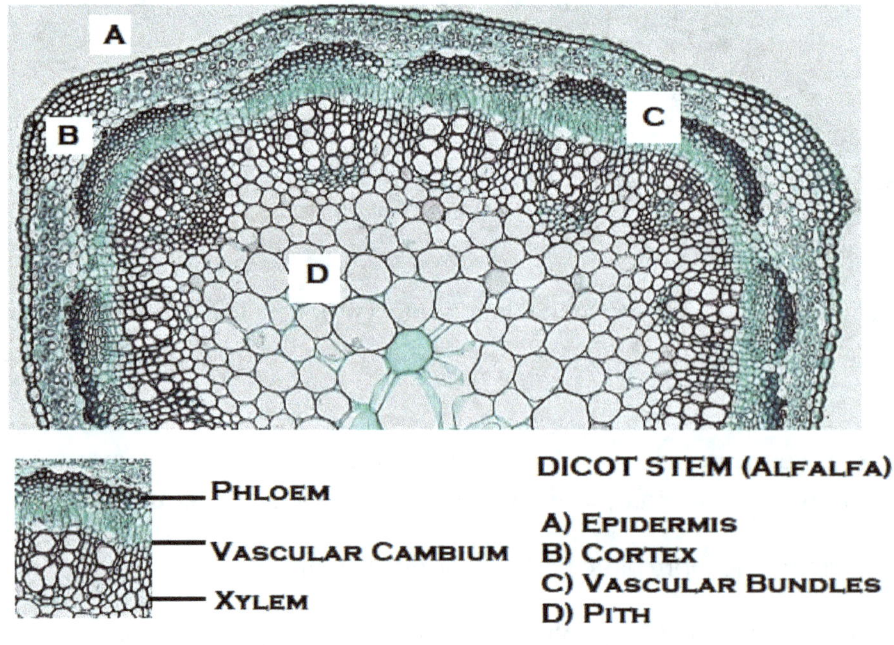

34. As you can see, the vascular bundles in a dicot stem form a concentric pattern under the cortex layer. The **vascular cambium** meristematic layer divides to produce phloem toward the outside and xylem to the interior.

35. Both the cortex and the pith are layers of ground tissue that store water and starch. The cortex is more likely to be impregnated with more secondary compounds than the cells toward the center pith.

36. Now let's consider what happens in the many species of dicots that produce woody growth and become shrubs, bushes, trees, or woody vines when they mature. Their stems undergo significant anatomical changes.

37. **Dicots** are capable of forming true woody tissue by layering the **xylem** in concentric rings.

38. In general, it takes 3 growing seasons for a dicot plant to grow true wood. At first, sapling stems look just like the stems of herbaceous dicots, but as they mature, tissue types change in prominence.

39. Woody growth is the result of cell division from both concentric rings of meristematic cambium tissue.

40. Toward the outside, cells in the **cork cambium** divide to create a specialized layer of outer bark known as the **periderm**. Periderm tissue is a specialized type of ground tissue impregnated with fibers and compounds.

41. As bark develops in **dicot** trees and shrubs, the **dermal tissue** splits and sloughs off, as it is pushed out and eliminated by growing periderm tissues from a the **cork cambium**.

42. You can see evidence of this happening in older saplings, where the growing trunk still has areas of dermis interspersed with layers of developing bark. In most tree species, the green dermis is gone by year four.

43. Inside the trunk, a deeper meristem layer called the **vascular cambium** undergoes mitosis and divides in both directions, laterally and medially. Phloem and xylem form from the vascular cambium.

44. Daughter cells on the outside (toward the bark) become the **conducting tissue (phloem)** that carries the sap in a tree trunk, while the inner cells (toward the heartwood) die off to become rings of wood (**xylem**).

45. Each year, both the cork cambium and vascular cambium undergo divisions duringthe spring and summer growth seasons, laying down extra concentric layers of both types of tissues.

46. **Primary phloem** emerges from the **procambium** layer of the shoot tip, the **apical meristem** of the tree. It has long and thin sieve tubes, is fibrous toward the outside, and has few parenchyma cells in between its vessels.

47. **Secondary phloem** emerges from the **vascular cambium** each season and adds layers to the inner bark year-after-year. Only the innermost layers of the phloem are very useful for conducting sugars.

48. In the process of growing into a tree, the phloem layers eventually crush and destroy the cortex ground tissue.

49. Older layers of inactive secondary phloem die and pile up, fusing with the outer cork layer of the bark and becomes almost indistinguishable in many types of trees.

50. Similarly, **primary xylem** emerges toward the **apical meristem** as the tree grows taller, while **secondary xylem** are the familiar tree rings that grow during the spring and summer of each growing season.

51. As xylem moves toward the center of the tree each year, it gets crushed and compressed and stops conducting water. This eliminates the **pith** as the tree grows.   **Heartwood** becomes the primary support beam of the tree.

52. Most water conduction in trees is done by the outer xylem layers, and little to none is done in the center.

53. This is why it doesn't hurt a big oak tree to have some sort of vermin like a possum or raccoon move into a hollow cavity at the center...it was dead anyway. NEVER stick your head in a hollow tree, by the way.

54. However, a thunderstorm can have catastrophic consequences, since their main support structure disintegrates and allows tons of branches, leaves, and rotten wood to come crashing down on the new Jeep in your driveway.

55. The age of a tree, as you probably know, can be determined by counting EITHER the light rings or dark rings and adding three years to account for the juvenile stage when the stem was still herbaceous.

56. Why light or dark and not both? **Spring wood** is lighter because the xylem are more loosely packed, while **summer wood** has more densely packed xylem, since growth slows down when it is hotter and drier.

57. Historical rainfall, year-by-year, can also be analyzed by looking at the width of particular rings. In wetter years, there is more xylem growth, while seasons of drought produce thinner rings.

58. The diagram below depicts the macro-anatomy of a typical woody stem. In this case, this is a young shrub, so there is still a layer of pith (ground tissue) at the center. This will disappear as the shrub adds xylem layers.

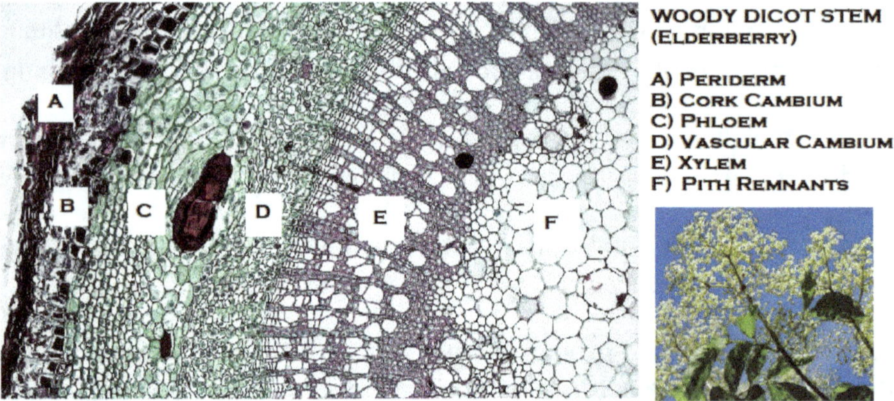

59. In addition to the concentric inner meristems of the cambium, dicot stems also have **apical meristems** called **terminal buds** at the shoot tip and **lateral meristems** called **axial buds** on the sides of the trunk.

60. Apical meristems are involved in vertical growth, sprouting new crowns of branches and leaves, while lateral meristems divide to produce new branches and leaves.

61. Lateral meristems are obvious on the sides of the stem, since growth emerges from knotty areas called **nodes**, while bare smooth bark or dermis between these dividing tissues are known as **internodes**.

62. Dicot stems also develop scars as leaves and buds shed with each passing season.

63. **Leaf scars** from prior years are circular indentations on the sides of branches. **Bundle scars** of vessels can be seen at the center of leaf scars as pinholes where xylem and phloem pipes once connected stem and leaf.

64. **Bud scale scars** develop as concentric rings around the growing shoot tip. These can be counted to determine the age of individual branches, since the apical bud drops off each year during fall or winter.

65. While stomata are not visible to the naked eye, **lenticels** can be seen. These are pinhole-like opening in the bark that allow gas exchange to occur. These are necessary to keep the phloem layer underneath alive.

66. Each of these features is labeled on the branch pictured below.

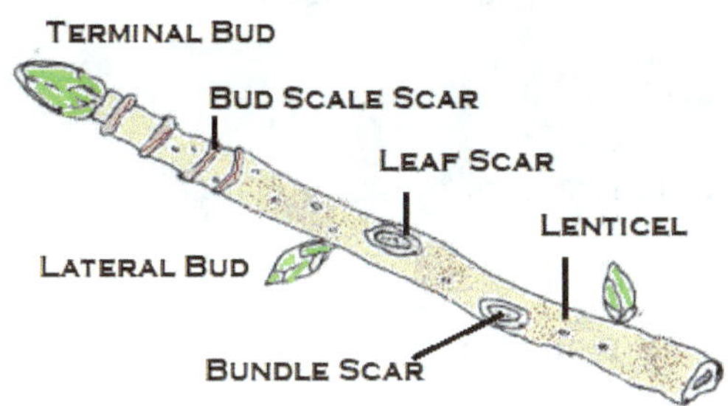

C) Root Anatomy and Histology

1. **Roots** have several basic primary functions, including anchoring plants in the soil, absorbing water and dissolved minerals, and storage of surplus carbohydrates in the majority of plant species.

2. All roots develop from the embryonic **radicle** and form an **apical meristem** at the tip called the **root cap**.

3. Monocot roots tend not to develop **lateral meristems** other than the vascular cambium, but dicot roots develop lateral meristems that allow the main **taproot** to branch out into smaller side roots.

4. With the exception of **epiphytes**, such as certain orchids that grow aerial roots to cling to rainforest trees, there are relatively few plants with photosynthetic tissues in the roots. Obviously, it's dark underground.

5. **Monocot roots** tend to be small and fibrous, while **dicot roots** often rely on a large central **taproot** that penetrates the soil deeply and sprouts smaller roots from **lateral meristems**.

6. The shape of roots are also heavily influenced by plants' evolutionary adaptations to environmental pressures.

7. The table below depicts some of the many different types of roots found in nature.

8. While there are distinct differences between the roots of **monocots** and **dicots**, there are still many common functions in their tissues.

9. The outer **epidermis** layer covers the root of the plant and absorbs water and minerals from the soil. There are almost always tiny extensions of the epidermis called **root hairs**.

10. **Root hairs** increase the surface area of roots by hundreds to thousands of times.

11. Additionally, many plants have symbiotic **mycorrhizal fungi** growing on the roots that also increases the water-absorbing capacity of roots dramatically.

12. Inside the dermis is the **cortex**, where sugars and water are stored, and where the plant may make secondary compounds to prevent it from being consumed by burrowing animals.

13. Did you know, for instance, that raw potatoes are toxic?

14. The **cortex** contains large spherical **parenchyma cells** packed out with starch and water. There are abundanta spaces between the cells.

15. The parenchyma cells are also loaded with **plasmodesmata** that allow materials to easily flow through them and make their way to the **vascular tissue** at the center of the root.

16. The next layer in is the **stele**, which is sealed off from the outer part of the root with a thick layer of wax excreted by an outer ring of cells called the **endodermis**. The stele contains all of the vascular tissue.

17. Some roots also have a starchy **pith** made of ground tissue at the center of the root.

18. At this point, it is useful to go ahead and talk about the differences between monocot and dicot roots, as the structures from the **stele** inward vary 19. the two groups.

19. **Monocot roots**, as mentioned, tend to be thin and fibrous and do not penetrate the soil as deeply.

20. Externally, they are surrounded by the **epidermis**, which extends into smaller **root hairs** to increase the surface area of water and nutrient uptake.

21. The **root cap** serves as the **apical meristem** at the tip of the root. This is the center of **primary growth** that lengthens the roots and pushes them deeper into the soil.

22. The next layer in is ground tissue, the **cortex**, which is one of the main two sites for food storage and for the manufacturing and secretion of secondary compounds that discourage herbivores, such as in horseradishes.

23. At the end of the cortex, a layer of cells called the **endodermis** separates the vascular tissue from the cortex. It does this by secreting a layer of wax called the **Casparian strip** on its inner surface.

24. Water is free to flow toward the center of the root, but the Casparian strip prevents backflow out of the root.

25. Water flows easily from the epidermis to the cortex and then through the **endodermis** because of **capillary action** and **water potential** differences. Big spaces between the fluffy round cortex cells help.

26. Tunnels called **apoplasts**, which are specifically designed to serve as water pipes run all the way through the cells of the cortex (created by **plasmodesmata**) to the xylem in the center.

27. If water goes AROUND the cells, there is another system of tunnels created by empty spaces called **symplasts**.

28. Inside of the **endodermis**, there is a concentric ring of **meristem tissue** called a **pericycle**. It is indistinguishable from the rest of the cells until it begins to divide to make lumps. The **pericycle** divides to produce lateral roots.

29. Interior to the endodermis, **phloem** forms another concentric ring, which is broken up by **xylem**

30. Finally, at the center of the root is a starchy **pith** made of ground tissue (parenchyma).

31. Scattered throughout the pith and the cortex, there can also be **collenchyma** and **sclerenchyma** that toughen up the root and prevent it from being chomped to pieces by moles.

32. The pictures below show cross-sections of a typical monocot root.

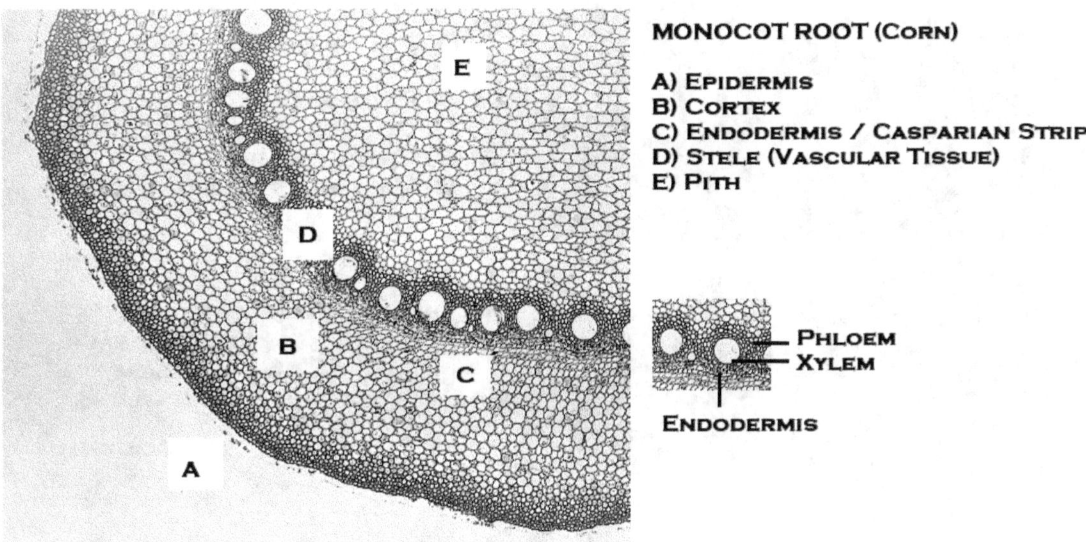

33. As previously mentioned, it is more typical for **dicot roots** to be large taproots. These large central taproots then sprout smaller side roots from **lateral meristems.** A carrot is a textbook example of a dicot root.

34. Dicot roots in woody plants also gain **secondary growth** to create wood fiber, but they lose their absorptive function and just serve as anchoring and support structures at that point.

35. Roots can become woody in these plants, because the **vascular cambium** layer of **meristem** extends down into the roots and joins up with the **pericycle.**

36. The **cork cambium** also extends into the root and divides, crowding out the epidermis, cortex, and the endodermis of a mature woody root.

37. However, even woody plants MUST have soft pliable absorptive roots deeper in the soil.

38. **Dicot roots** are similar to **monocot roots,** in that they are surrounded by an epidermal layer that grows from the root cap. However, dicot roots tend to have many more **root hairs.**

39. Since dicot roots aren't fibrous, this helps them to increase their surface area dramatically.

40. **Dicot plants** also commonly have **mycorrhizae** and symbiotic nitrogen cycle bacteria in and on their roots.

41. Internally, the first layer of a dicot root is also the **cortex**, which also gives way to the **endodermis** and the **pericycle.** From there, things are different.

42. The **vascular cambium** in **dicot** roots continues to divide and lay down layers, as it does in a growing tree. As a result, a dicot plant loses its pith, as the vascular tissue crushes it.

43. Dicot roots have a distinct layer of **xylem** in an 'X' shape inside the pericycle.

44. The **phloem layer** sits under the arms of the 'X'.

45. The pictures below depict typical dicot root cross-sections.

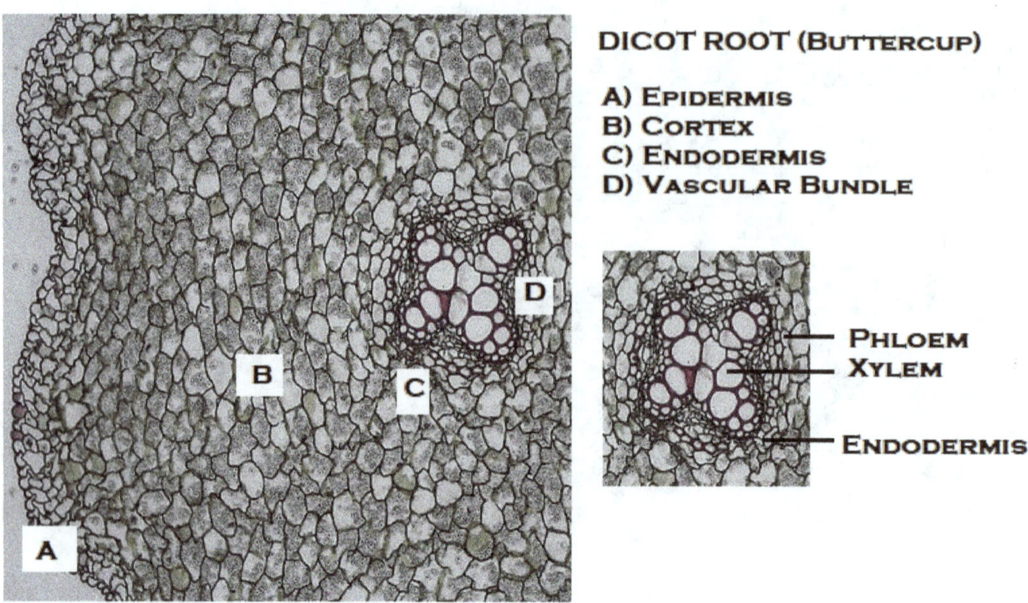

46. We will wrap up this chapter with another moment from my career that is reflective of the extreme degree of randomness that is my life. Somehow we won the next game, in spite of a completely worthless practice.

Circa 2012

We tried....We really DID! Just like many high schools, our boys and girls soccer programs had to share a field. Fortunately for us, there was an old municipal stadium from a bygone era. When it was the girls' turn to have the stadium, we would take a bus over there. In this particular season, we were trying to get the last playoff spot in our region and we practiced on a Friday afternoon bye. Coincidentally, the practice field was adjacent to the National Guard Armory, who happened to be hosting a circus for charity...THAT FRIDAY about 3 hours later. The first interruption came when an 18-wheeler pulled up sounding like were individually offloaded as our players ignored their small-sided game. Five minutes later, it was time for the elephants to practice. Once again, practice came to a screeching halt. Another 10 minutes and we got our concentration back and were in the middle of 3-man weave. That's when a four foot tall clown decided to practice popping wheelies on a moped for his later performance. Exasparated, the other coach and I managed to ignore this as well. The death knell for our practice? Two camels had a heated disagreement. Camel #1 pinned the other to the ground and repeatedly heat-butted him while camel #2 tried to gnaw his leg off. I will never forget the demonic honking and screaming. With that, coach and I gave up and bet a meal on who would win..... Neither did. A clown with a push broom beat them both senseless until they learned some manners.

Chapter Seven

Plant Growth and Physiology

S ECTION 1: Translocation of Materials through Plant Tissues

A) Transpiration and Water Transport

1.Now that you have a command of how plants are classified and how they are anatomically built, we now turn our attention to how homeostatic processes work on the cellular level in plants.

2. Transport of materials through the **vascular system** of plants is one of the most fundamentally critical set of processes necessary to keep plants alive in a relatively immediate time frame.

3. Let's begin with looking at the most basic of needs, the need for water. **Water translocation** occurs in the hollow **xylem** tissues of vascular bundles and follows a few basic laws of physics.

4. **Capillary action** draws water into the narrow spaces in the hollow centers of **vessel elements** and **tracheids**.

5. The forces of **cohesion** and **adhesion**, collectively known as capillary action, are stronger than gravity when applied over a narrow diameter. This same phenomenon is why coffee climbs up a coffee stirrer in a drink.

6. **Cohesion** refers to water hydrogen bonding to other water molecules to form a lattice that creates droplets.

7. **Adhesion** refers to water hydrogen-bonding to other molecules. In this case, water forms hydrogen bonds to oxygen atoms sticking off of the sugars of cellulose fibers. Water droplets form and stick to these fibers.

8. As hydrogen bonds form, more molecules keep climbing the column, attracted to other polar molecules above their location. This causes the water to keep climbing up the xylem.

9. There is a second fundamental force that sucks water up a tree trunk. This is due to the **vapor pressure differential** between the inside of the tree trunk and the surrounding air outside of the tree.

10. Since the water molecules are being forced through miniscule openings in the xylem, the water in the trunk is under high pressure inside the xylem compared to the surrounding vapor pressure in the air outside.

11. Consequently, when the tree's leaves opens their **stomata** to take up CO_2 for photosynthesis, the high pressure inside the plant pushes water vapor out into the surrounding atmosphere.

12. This creates negative pressure suction that pulls the whole column of water molecules up the xylem vessels in the area near the open stomata. It is the same principle as sucking a milkshake through a straw.

13. The **transpiration-cohesion model** combines these two principles to describe the mechanism that moves water molecules from the roots of a tree and up to the stem and the leaves.

14. The diagrams below depicts the transpiration-cohesion model.

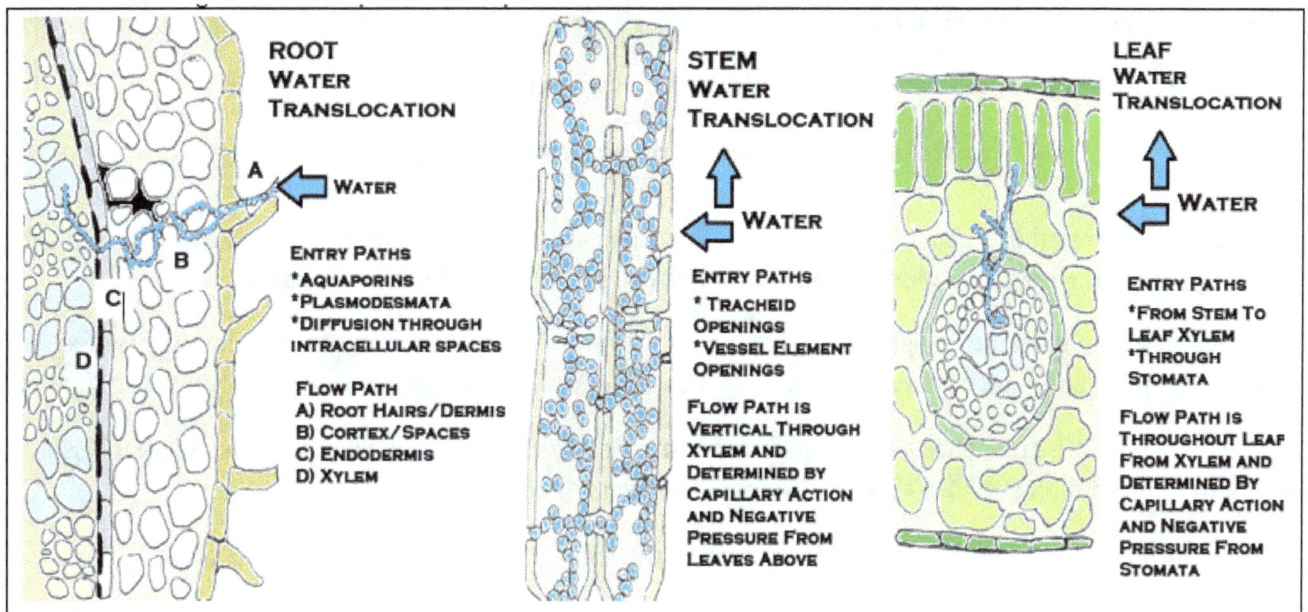

15. As you can see, water first makes its way through the roots. Starting at the epidermis, **osmosis** and **capillary action** causes the water to squeeze through intracellular spaces and **aquaporins** to move **to** deeper layers.

16. Since the **cortex** is higher in carbohydrate concentration and dissolved nutrients, water will naturally flow into those cells by **passive transport**. This continues until the water enters the **endodermis**.

17. Water is still free to move forward from the endodermis to the **vascular tissue**, but the waxy **Casparian strip** on the interior surface of the endodermis prevents backflow. From there, water enters the root **xylem**.

18. A combination of capillary action and negative pressure drags the column of water all the way up the stem and into the **vascular bundles** of the leaves.

19. Eventually, the unused water molecules exit the tree via the billions of leaf **stomata** in the canopy.

20. This phenomenon is known as **transpiration**. There is such a large quantity of water moving through trees that transpiration has significant influences on local temperature and weather.

21. This is why forests are cooler than open sunny areas. Besides shading, the other factor that creates natural air-conditioning is evaporating water from transpiration. Water can carry away significant amounts of heat.

22. As the evaporated water rises in the atmosphere is undergoes **condensation** into clouds. Trees influence local weather cycles in forested areas, and they are the reason for daily thunderstorms in **tropical rainforests**.

23. It is possible to predict the direction of water flow through a tree and enumerate the strength of the pressure behind the flow by mathematically determining the **water potential**.

24. **Water potential** is a physical value equal to solute pressure plus atmospheric pressure. Regardless of whether a system is living or not, fluid flow behavior is dependent upon two factors.

25. Those factors are **atmospheric pressure** and **solute pressure** due to osmotic flow.

26. The formula for water potential is shown below:

$$\Psi_{water} = \Psi_{solute} + \Psi_{atmospheric\ pressure}$$

$$\text{Where } \Psi_{solute} = -iCRT \quad \text{(ionization constant)(solute concentration)}$$

$$\text{(molarity-pressure constant)(temperature Kelvin)}$$

27. When water potential (Ψ) has a positive value, that means water will create pressure to move away from a source. When Ψ is a negative value, that means water will move into a source via suction.

28. Since the solute pressure is higher inside of a plant's xylem, due to a column of water forcing itself against the walls, the pressure will be lower outside. This means the water is going to leave if the stomata are opened.

29. The more water the plant has stored, the faster **transpiration** will go, as well, due to the fact that dissolved solutes are more dilute, and the solutes exert less osmotic pressure.

30. **Atmospheric pressure** differences exert the greatest effect on the movement of water through trees.

31. **Solute potential**, however, also accounts for part of the vertical movement of water in plant vessels.

32. This is because plants have large sinks of starch and sugars that cause the roots and stems to have a lower **water potential** than the soil. Therefore, water moves towards the dissolved substances.

33. Since the relative concentration of carbohydrates inside the plant tissues are dramatically higher than their concentration in soil water, the reverse is true of water. The concentration of water in the cells is less.

34. Since water will follow its concentration gradient, it will flow from the source (100% water or close to it) in the soil, to the sink (70% water) in the cells of the plant. Solutes are not free to move, so they don't factor.

35. **Solute concentration** has a strong influence on fluid flow, since more dissolved substances means lower concentrations of water in those cells. In turn, this will have a stronger pull on incoming water.

36. The **ionization constant** refers to how well particular solutes will dissolve. Substances that dissolve completely have a much stronger contribution to osmotic pressure than partially or poorly soluble substances.

37. Osmotic forces are the initial influences that draw water into roots.

38. Once the water seeps through and enters the **xylem** in the roots, it is then pulled away by **cohesion** and **adhesion**, creating upward suction inside the root and on into the stem.

39. In turn, the vacuum left behind by moving water molecules will pull more water molecules in to fill the space left behind as the choo-choo train of water molecules climb up the tree.

B) Translocation of Organic Molecules

1. **Organic materials**, primarily **carbohydrates**, must be transferred from the photosynthetic mesophyll cells to all the other cells in the plant, feeding them so they can run **cellular respiration** and fill leucoplasts with food.

2. Ultimately, all the biological molecules produced by the plant are spin-offs of photosynthetic products. The initial **G3P molecules** produced from the **Calvin-Benson cycle** can be modified into many different compounds.

3. Most commonly, plants will move simple sugars, such as **fructose** and **glucose** and modify them later in ground tissue cells. Other metabolic pathways also produce **amino acids** and **lipid** precursors that are also moved.

4. Sugars and other organic molecules enter the plant's **sieve tube elements** within the phloem vessels, with help from **bundle sheath cells** and the phloem **companion cells**.

5. Depending on the concentration of the sugars at the time, glucose and fructose either ender the bundle sheath cells through **carrier proteins** by **passive transport** or they must be pumped by **active transport.**

6. The same process repeats itself as sugars move from the bundle sheath cells to the companion cells.

7. Once the sugars arrive at the companion cells, they can easily move the sugars into cavernous spaces of the **sieve tube elements** through the numerous enormous **plasmodesmata** that join the partners together.

8. Since the **xylem** are right next door to the **phloem** cells, sugar transport also gets a boost from **osmosis.**

9. Xylem is also full of holes, allowing water to seep freely into the phloem layer. Since the concentration of dissolved substances in the phloem is dramatically higher than the xylem, water will be drawn into the phloem.

10. Now that sugars are dissolved in the inflow of water coming from the xylem, a thick syrupy liquid (sap) begins to form and move downward. Because of negative **solute pressure**, this will continue all the way to the roots.

11. At the bottom of the tree, nearly pure fresh water (with some dissolved minerals) is coming up from the soil.

12. **Osmosis** will cause freshwater is going to want to move up, while the sugars are going to want to move down, so that both **concentration gradients** equalize.

13. The concentrated sugar solution will keep moving down until it equalizes in **osmotic pressure.** This won't happen until it gets somewhere far into the stem (such as a potato) or deep in the roots (such as a carrot).

14. Along the way, all of the plant's cells take their cut and remove sugars from the phloem for metabolism, much like the cells of animals remove the nutrients that they need from the circulatory system as plasma passes by.

15. This slows the flow rate of the fluid down a little, since it raises the **water potential** of the sap as it loses sugar concentration. However, what the sap loses in speed from concentration, it gains back with increasing gravity.

16. Therefore, it is no problem to tap into the phloem of a maple trunk and get a bucket full of syrup. At least that's half the story anyway. In reality, maple sap is about 30 times less concentrated than syrup and must be boiled.

17. So that little 16 ounce bottle of maple syrup in your kitchen cabinet really took nearly four gallons of sap to produce. This is why the real thing costs significantly more than the disgusting brown corn syrup swill eaten by the unenlightened and unwashed.

18. The diagram below depicts the movement of sugars through a tree trunk.

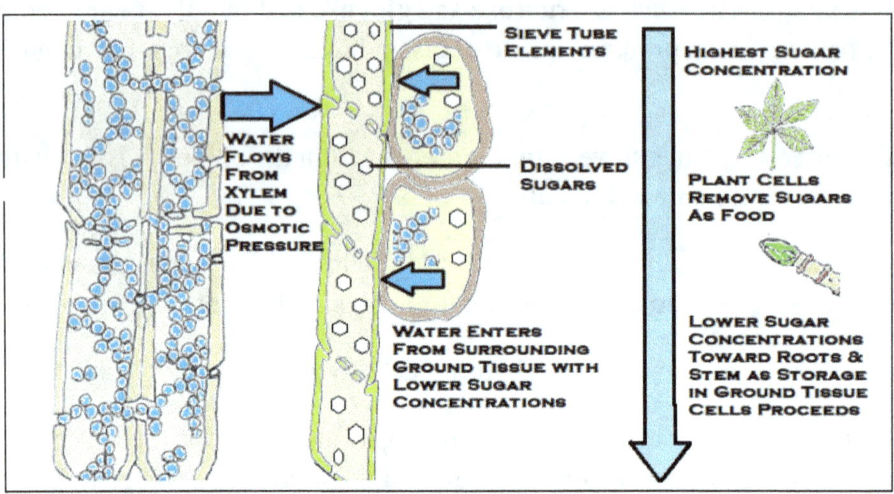

SECTION 2: Growth Responses in Plants

A) Factors that Influence Growth: Hormones and Tropisms

1.Plant growth, somewhat obviously, is regulated by many of the same factors that determine the rate of photosynthesis, and hence, **primary productivity**.

2. Among the factors that regulate growth are solar energy, temperature, water, availability of soil nutrients, presence of symbiotic microorganisms, and oxygen levels.

3. One response to these factors is the manufacturing and distribution of **plant hormones**, which signal cells within various tissues to make changes within the plant.

4. Though chemically different from the hormones in animals, the same end goal is achieved.

5. As you may remember, **hormones** are biochemical signals that act one of two ways to cause DNA transcription, mRNA translation, and the subsequent manufacturing of a needed protein.

6. Some hormones trigger **receptors** on the cell membrane surface, causing **secondary messengers** within the cell to phosphorylate (and turn on) proteins that influence **transcription**.

7. Other hormones diffuse directly through the membrane and directly turn on **transcription factor proteins** on the DNA. Either way, changes occur in the cell affected by the hormone.

10. Plants have the added problem of cell walls, which can be quite thick, so membrane **receptors** are less useful than they are in animal cells.

11. As a rule, hormone molecules used by plants need to be very small to allow them to pass through the spaces between the fibers of cell walls. They are often modified from simple biochemical monomers.

12. The table below lists the major classes of hormones in plants, describes their molecular structure, and gives their sources of origination and their targets.

Hormone	Produced In	Effects	Molecular Structure
Auxins	Shoot meristems Young leaves Seeds	Cell Elongation Stem Elongation Apical Dominance (Plant getting taller instead of bushier) Root Growth Fruit Development & ripening Stimulation of Ethylene Production	Derived from the amino acid tryptophan. Side chains are changed to alter the target and effects of different auxins.
Gibberellins	Young leaves Shoot meristems Embryo of seeds	Seed germination in response to light Stem Elongation Flowering in response to light Fruit Development Delays death of flowers and fruit Stimulates enzymes that mobilize sugars	Derived from acetyl-Co A in metabolism. Long series of reactions makes these.
Cytokinins	Roots Meristems	Cell division Delay of leaf senescence (leaf dropping) Inhibition of apical dominance (promotes bushiness) Leaf expansion Chloroplast development Flower development Embryo development Seed germination	Derived from the nucleotide adenine. Side chain changes allow for more than 200 different types.
Ethylene gas	Stem nodes Ripe fruit Damaged tissues	Fruit ripening Response to stresses Seed germination Root initiation Senescence and abscission of leaves & flowers (dying and dropping)	Synthesized from the breakdown of the amino acid methionine.
Abscisic Acid	Leaves and any other cell with plastids.	Seed dormancy Water stress responses Leaf dropping	Produced in chloroplasts from the metabolism of carotenoids.
Brassinosteroids	All tissues seem to produce some level of these.	Enhance the effects of other hormones Trigger metabolic changes Xylem formation Light-directed changes Cell division Stem elongation Root inhibition Flower development Leaf senescence (dropping)	Derived from cholesterol and other sterols, which are themselves made from Kreb's cycle precursors.

13. Plants have to adapt and adjust constantly to changes in the environment. Rain, drought, changes in solar energy levels, changes in soil chemistry, herbivore damage, and competition from other plants are such factors.

14. In turn, these environmental changes can trigger genetic changes. For instance, a hot dry period may cause plants to manufacture **heat shock proteins**. In turn, these go to other cells and trigger receptors.

15. Since leaves are a liability during a drought, the heat shock proteins trigger receptors that then cause the gene for the enzymes that manufacture **abscisic acid**. From there, the abscisic acid causes leaves to drop.

16. Plants can also make temporary adjustments by inflating and deflating cells with water.

17. This works, because plant cells can pump K^+ ions into our out of cells, causing water to follow or leave because osmosis. The resultant swelling or shrinkage can alter the shapes of various structures as needed.

18. For instance, **guard cells** open and close stomata under control of potassium pumps.

19. When there is sufficient water, plants will open their stomata to take up CO_2 for the **Calvin-Benson cycle** of photosynthesis. The tradeoff, of course, is that open stomata lose huge amounts of water vapor.

20. The **stomata** open, because K^+ ions are pumped into the guard cells, causing water to follow osmotically. Tethered together at their ends, the guard cells swell and bow out from each other, opening the stomata.

21. When it becomes too hot and dry to be worth the cost of taking up CO_2, the plant closes the stomata by pumping the K^+ ions back out of the cells. The water leaves and the guard cells close.

22. The diagram below summarizes the mechanism used to open and close guard cells.

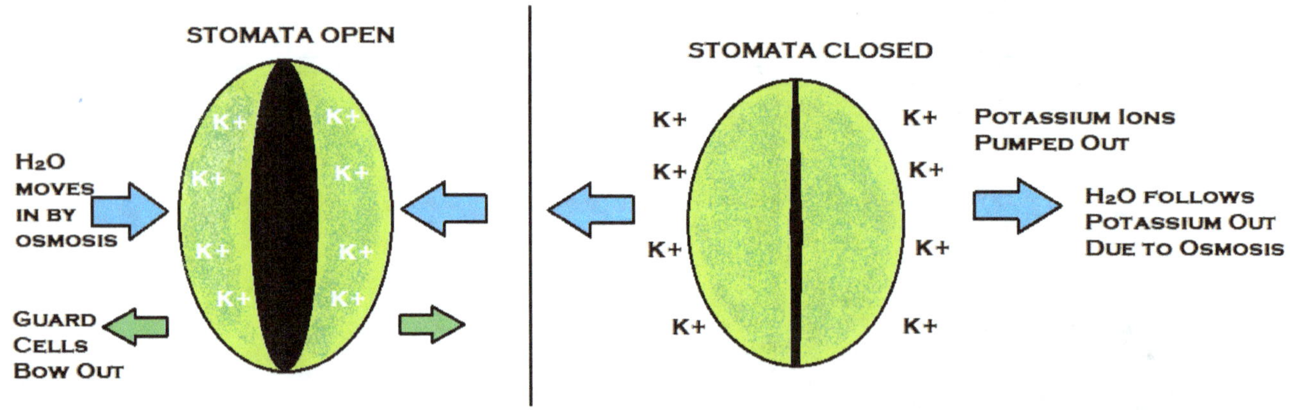

22. Besides guard cells, this system can be used on a larger plant-wide scale.

23. For instance, rhododendron leaves close in response to stresses, such as drought, cold, damage, or pressure. This protects leaf surface from freezing or drying, or from snow piling onto leaves and breaking the branch.

24. Mimosas close their leaves in response to touch to discourage herbivores from grazing on them.

25. Also mimosas (and shamrock plants too) fold up their leaves in response to falling light levels, closing at night.

26. Sunflowers, if you've never noticed, change pressure in various regions of their stems, allowing them to follow the sun in the sky, changing the position of their stem to gather the most light at different times of day.

27. All of these are examples of plant **tropisms**, which are physiological plant responses to some immediate environmental condition. Changes are triggered by each stimulus.

28. **Tropisms** are often regulated by hormones. Different stimuli active different hormones and different pathways.

29. The table below lists the major known tropisms that plants exhibit, with descriptions and examples. Below the table, is a picture explaining how auxins cause phototropism.

PHOTOTROPISM PLANTS GROW AND BEND TOWARDS A LIGHT SOURCE.	IN RESPONSE TO LIGHT, MORE AUXINS AND GIBERELLINS CAUSE ONE SIDE TO GROW FASTER. EXAMPLE: PLANT BENDS TOWARD LIGHT AT WINDOWSILL.	
HELIOTROPISM PLANTS GROW AND BEND TOWARD THE POSITION OF THE SUN IN THE SKY.	IN RESPONSE TO POSITION OF SUN, PLANT CAN USE ION CHANNELS TO CHANGE OSMOTIC POTENTIAL, INFLATING OR DEFLATING STEM WITH WATER. EXAMPLE: SUNFLOWERS WILL ADJUST ANGLE TOWARD SUN.	
CHEMOTROPISM PLANTS GROW ROOTS TOWARDS WATER OR NUTRIENTS AS THEY SENSE A CHEMICAL TRAIL OR MOISTURE.	POSITIVE STIMULUS (TOWARD) OR NEGATIVE STIMULUS (AWAY) CAUSES HORMONAL CHANGES THAT RESULT IN FASTER OR SLOWER LOCAL GROWTH. EXAMPLE: TREES GROW ROOTS TOWARD A STREAM.	
GRAVITROPISM PLANTS SENSE GRAVITY AND REMAIN UPRIGHT, WITH SHOOTS GROWING UP AND ROOTS DOWN.	INTRACELLULAR PARTICLES FALL TOWARD GRAVITY. THIS CAUSES A CASCADE OF SIGNALS THAT AFFECTS DIRECTION OF GROWTH. EXAMPLE: FALLEN TREE GROWS NEW UPRIGHT LIMBS.	
THIGMOTROPISM PLANTS TOUCH ANOTHER STRUCTURE AND ALTER THEIR GROWTH PATTERNS TO ACCOMODATE IT.	TOUCHING A SURFACE HORMONALLY STIMULATES DIFFERENTIAL GROWTH ON ONE SIDE OF PLANT. EXAMPLE: VINES CLIMB A TRELLIS.	

30. So, the action of hormones and the **tropisms** that they trigger are the primary means by which plants adjust to feedback from the environment and regulate their growth.

31. The diagram below explains the process of phototropism. Just about any plant, left long enough in an area with unequal solar energy, will begin to show **phototropism.**

32. In this case, solar energy triggers biochemical changes inside of the cells on each side of the plant. They send signals to the cells manufacturing **auxins** at the shoot tip.

33. From there, auxins migrate differentially to the lit side and shaded side of the plant. The cells on the shaded side receive more auxins, so they divide more rapidly than the cells on the inside of the plant.

34. Over time, this will cause the plant to bend toward the sunlight.

35. Other tropisms work similarly. A stimulus causes messages to be sent to other cells that manufacture plant hormones. From there, receptors are triggered, genes are turned on, and hormones spur changes.

36. The diagram below shows the basic idea of how phototropism occurs in a new plant shoot.

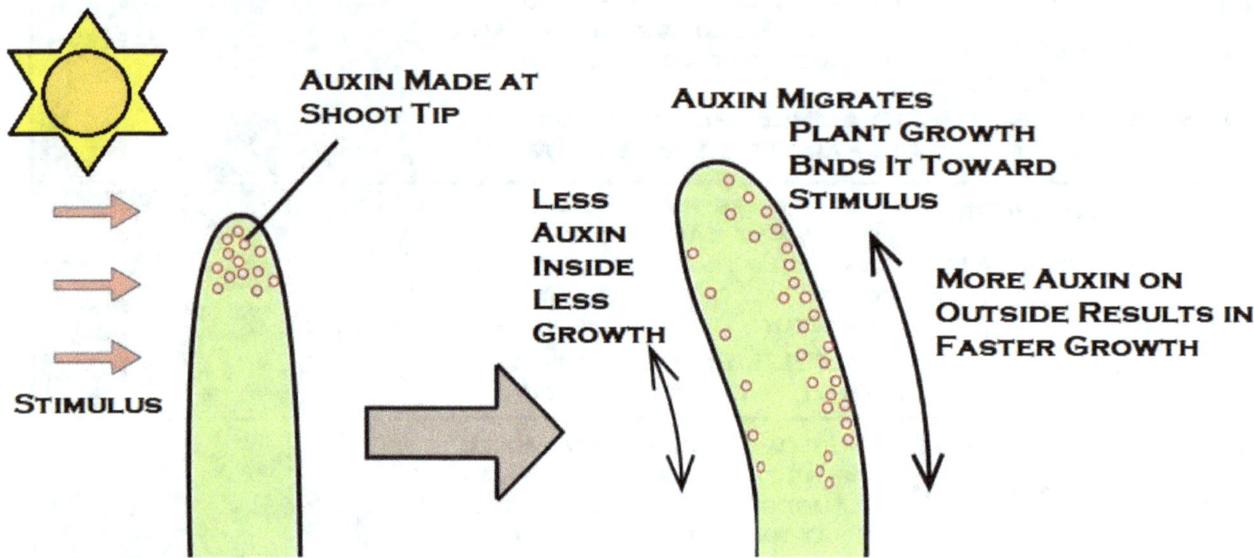

B) Plant Nutrition

1.Thousands of years of trial-and-error and formal agricultural research has gone into understanding plant nutrition, as it has a dramatic effect on the survival and productivity of all plant crops.

2. There are numerous components to understanding plant nutrition, but they all start with the soil.

3. All species of plants have optimal soil conditions that makes them most productive. Soil moisture, pH, organic content, particle size, **macronutrient** components, and **micronutrient** components all matter a great deal.

4. Furthermore, it is the interaction between all of these factors that determines the optimal habitat for a particular crop species or cultivar.

5. **Macronutrients** are the environmental components that plants need in great quantities to maintain **homeostasis** and to grow and reproduce. These make up a minimum of 0.05% of a plant's dry mass.

6. Obviously, photosynthesis calls for two such macronutrients. All plants must have large quantities of **water** and **carbon dioxide** or they cannot run the **light reactions** or the **Calvin-Benson cycle** of **photosynthesisis.**

7. However, there are other **limiting macronutrients** that are often the X-factors that determine how much growth occurs (and how fast) by their relative abundance or scarcity.

8. While water CAN be a limiting nutrient during a drought, CO_2 is really never limited.

9. The most common limiting **macronutrients** are those identified on fertilizer bags. **Nitrogen, Phosophorus,** and **Potassium** deficiencies are the most common culprits for stalled or slowed growth in plants.

10. This is because these three macronutrients have direct limiting effects on plant biochemistry at the cell level.

11. Let's start with nitrogen. Without a nitrogen source, plants can't produce amino acids or nucleotides. In turn, they can't make the proteins, DNA, RNA, NADH, or ATP that they need to build and run the cell.

12. Since plants are producers, they must make these molecules from scratch by funneling G3P products from photosynthesis into different biochemical pathways that make amino acids and nucleotides.

13. **Nitrogen** becomes available to plants in the soil as **nitrate (NO_3),** which is soluble in water and comes through the roots. This is the only form of nitrogen that plants can directly take up and use in their metabolic pathways.

14. To get from the gaseous N_2 in the air, to usable nitrates in the soil, **nitrogen fixation**, **nitrification**, and **ammonification** bacteria in the soil must act on behalf of the plant.

15. For this reason, many plants have symbiotic bacteria living on and in their roots. In particular, legumes have specialized **root nodules** to house colonies of these beneficial bacteria.

16. Nitrogen limitation has also caused carnivorous plants to evolve in swampy areas. While Venus fly traps, pitcher plants, and sundews do consume flies, roaches, and even mice, it is not a carbon source that they are after.

17. The soil in swamps is so nitrogen-poor because of anaerobic **denitrification bacteria** that the plant has to resort to dead, rotting animals as a nitrogen source. In essence, these plants make their own compost piles.

18. The first diagram below recap the nitrogen cycle, showing the role of each component bacterial type. The second diagram below shows a flow chart of the nitrogen utilization metabolic pathways that plants use.

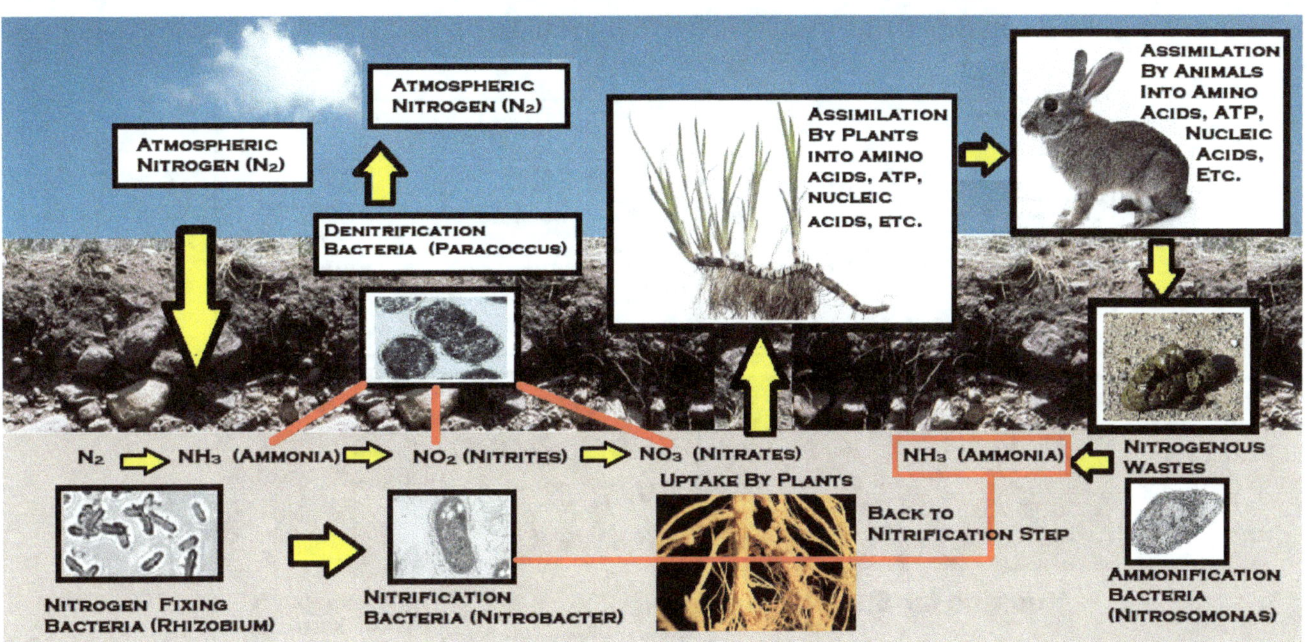

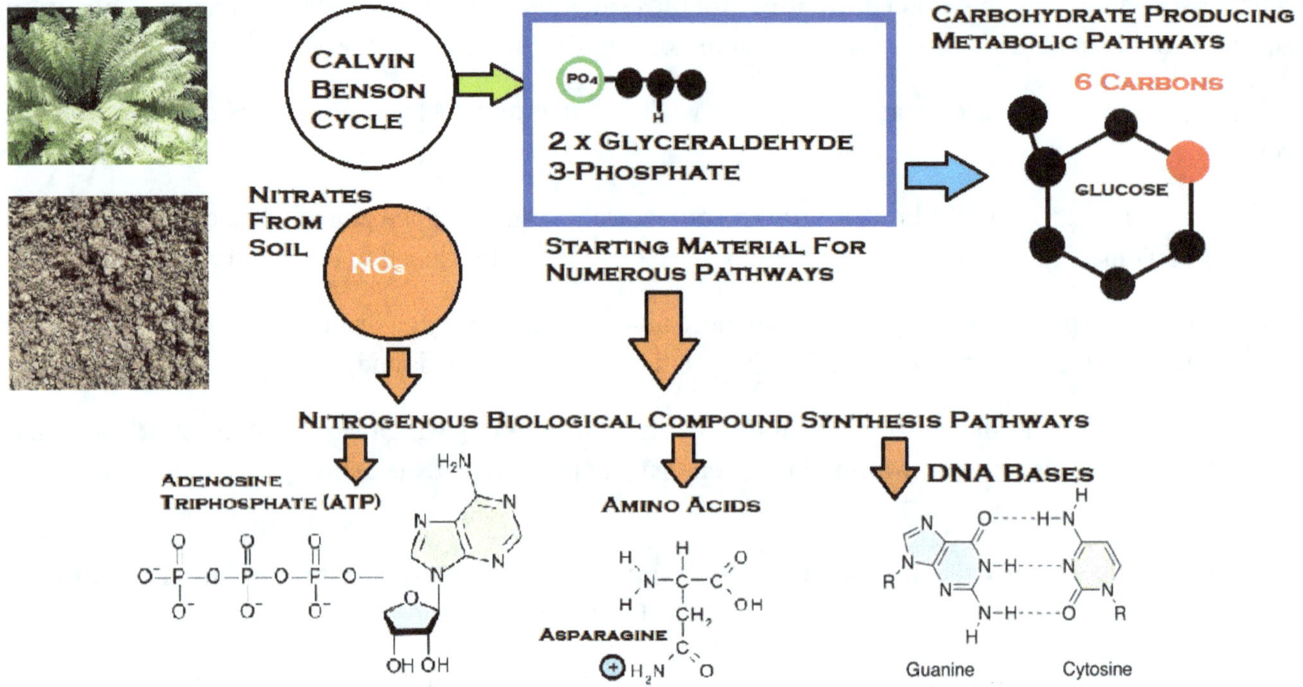

19. Another **macronutrient** needed by plants in large quantities is **phosophorus**. Phosphorus is a requirement for the manufacture of **phospholipids** for cell membranes and **nucleotides** for DNA, RNA, ATP, and NADPH.

20. Phosphorus is generally taken up by plant roots as soluble phosphate salts from the soil.

21. A third limiting macronutrient is **potassium**, which is needed in the membrane pumps of plant cells for active transport, and for keeping osmotic balance.

22. It shouldn't come as a surprise, then, that fertilizer bags say N-P-K on the side, as they are comprised of mixtures of these top three limiting nutrients.

23. Different plants species need N-P-K in different amounts at different times, so there are many formulas for fertilizers, such as general use 10-10-10, 16-4-4 for Spring leafing, or specific formulas for citrus and roses.

KOO-KOO FOR SHRUBBERY

Across	Down
1) LIRIOPE	1) LAUREL
5) LEYLAND	2) CHRYSANTHEMUM
9) HYACINTH	3) ROYAL
11) ELEPHANT EAR	4) HYDRANGEA
13) FORSYTHIA	6) LIGUSTRUM
14) WINDMILL	7) PYRACANTHA
	10) CREPE MYRTLES
	12) DAFFODIL

EATING THE PLANT KINGDOM

Across	Down
3) CABBAGE	1) AGAVE
5) OKRA	2) FIGS
7) PINEAPPLE	4) BREADFRUIT
8) PEPPERS	6) TAPIOCA
9) WILD RICE	10) CANTELOUPE
12) RHUBARB	11) PUMPKINS
14) MUSCADINE	13) RAMPS
16) SWEET POTATO	15) ROSE
20) LENTILS	17) PLANTAIN
	18) DURIAN
	19) BASIL

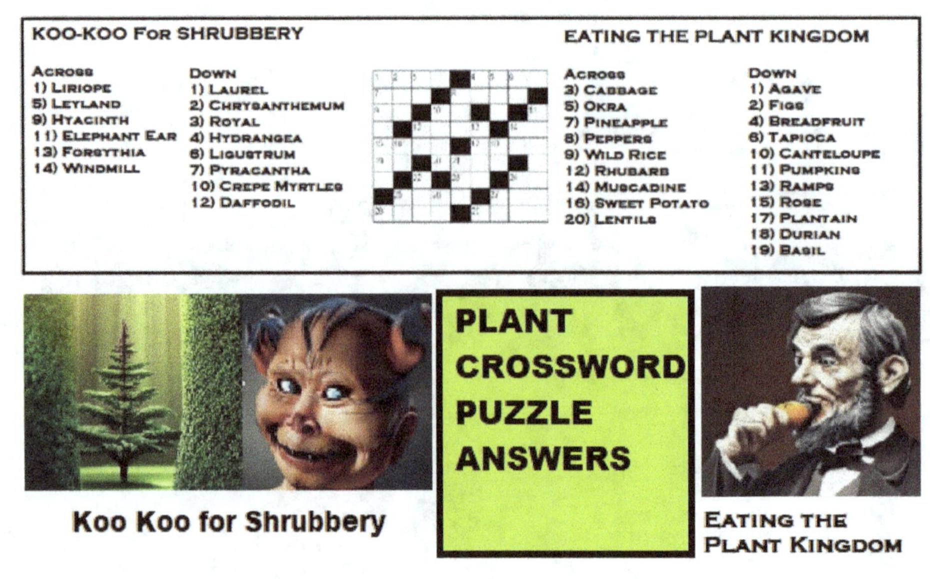

Koo Koo for Shrubbery   **EATING THE PLANT KINGDOM**

Macronutrient	Molecular Form	Purpose
Carbon	CO_2	Needed in carbs, lipids, proteins, nucleic acids. Incorporated during Calvin cycle of photosynthesis.
Nitrogen	NO_3, NH_4	Needed in proteins and nucleic acids and in NADH, ATP, chlorophyll, other protein coenzymes and electron carriers.
Hydrogen	H_2O	Taken up during light reactions. Needed to make carbs, lipids, proteins, nucleic acids and also needed for transport of water soluble nutrients.
Oxygen	H_2O, CO_2, O_2	Needed for aerobic respiration. Needed as a component of carbs, lipids, proteins, nucleic acids.
Potassium	K^+	Needed to keep osmotic and ionic balance. Needed for action of guard cells. Activates many different enzymes in plants.
Calcium	Ca^{2+}	Needed in cell walls. Affects membrane permeability. Activates enzymes. Serves as secondary messenger for hormones inside cells
Magnesium	Mg^{2+}	Found at the center of the chlorophyll molecule. Activates enzymes in carbohydrate metabolism.
Phosphorus	H_2PO_4, HPO_4	Needed in nucleic acids, phospholipids, ATP
Sulfur	SO_4	Needed in certain amino acids and vitamins
Silicon	SiO_3	Needed as a component of cell walls

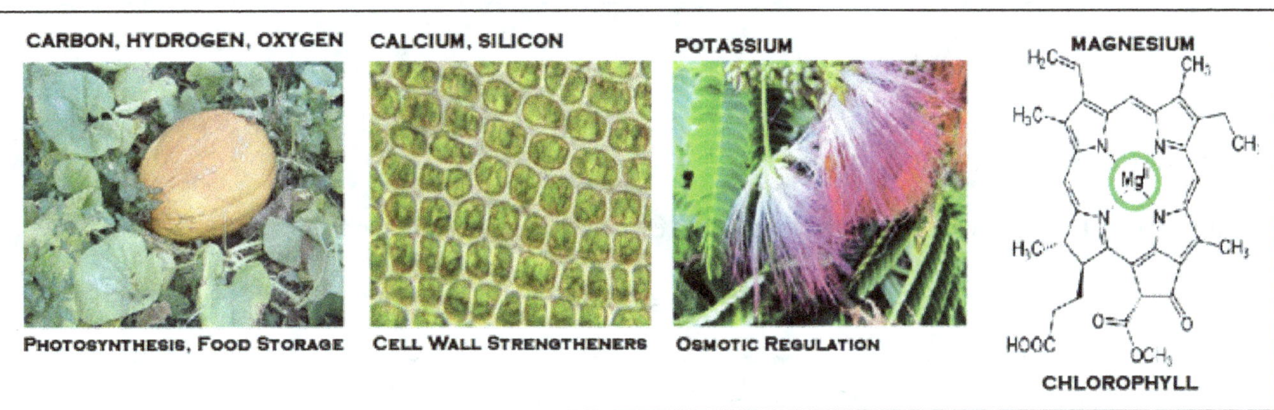

24. The first table above summarizes the major soil **macronutrients** and why they are needed.

25. **Micronutrients** are nutrients plants need in small quantities. Many of these can be toxic in excess quantities. Additionally, some of these micronutrients can also be relatively rare in the environment.

26. **Iron** is one of the most important of these micronutrients, because all heme proteins, such as the **cytochromes** used in both photosynthesis and cellular respiration, are built around iron-sulfur proteins.

27. Many micronutrients serve as **coenzymes**. A deficiency of one of these means that certain plant enzymes will either slow down or stop altogether, causing growth to also come to a screeching halt.

28. For instance, **manganese, zinc, nickel,** and **molybdenum** are coenzymes for many of the enzymes that convert nitrates into usable organic forms of nitrogen.

29. The second table and diagram below summarize the major soil **micronutrients**, along with their intended purpose.

Soil macronutrient	Typical form taken up	Purpose
Chlorine	C	Needed for ion balance. Enzymes in photosynthesis.
Iron	Fe2+ , Fe3+	Part of enzymes in the electron transport chains of photosynthesis, respiration, nitrogen fixation
Boron	H2BO3	Involved in active transport across membranes. Important in flowering and germination, fruit, and seed development.
Manganese	Mn2+	Involved in enzymes of respiration and nitrogen metabolism; needed in photosynthesis coenzymes
Sodium	Na+	Needed in photosynthesis. Needed to keep osmotic balance with K+
Zinc	Zn2+	Needed by enzymes in respiration and nitrogen metabolism
Copper	Cu+, Cu2+	Needed by photosynthetic enzymes
Nickel	Ni2+	Needed in nitrogen metabolism
Molybdenum	MoO42-	Needed in nitrogen metabolism

30. As you can see, most of these **micronutrients** are somehow involved in biochemical pathway enzymes that need **coenzymes** as activators.

31, Without these minor nutrients, photosynthesis or cellular respiration doesn't work.

32. The diagram below summarizes some of the biological functions of the micronutrients in the table.

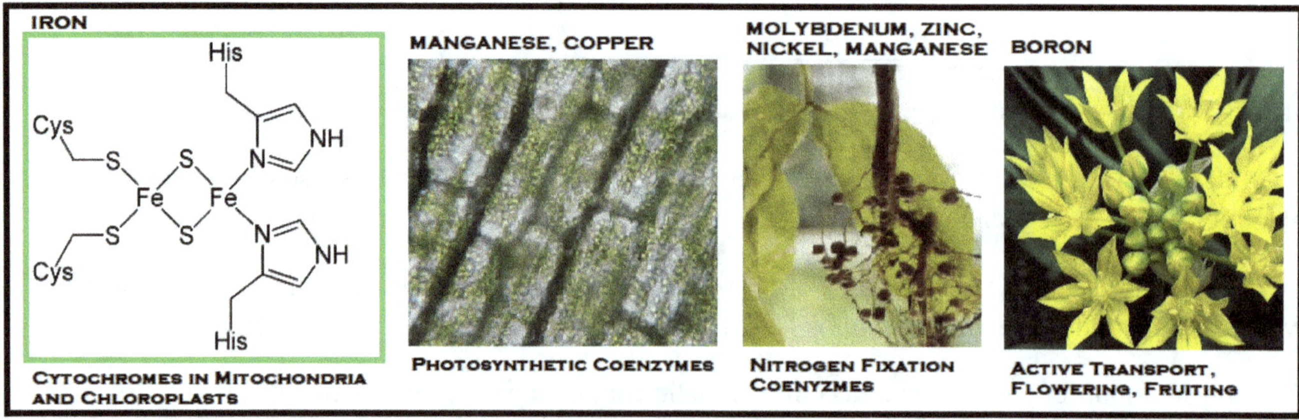

33. **Soil pH** greatly affects the solubility of certain minerals and the ability of plants to take up minerals.

34. Most plants grow best at a slightly acidic pH of around 6.0, since this is the 'sweet spot' where the most nutrients are soluble at the same time. However, this doesn't hold true for all species of plants.

35. It just so happens, that forest soil is slightly acidic from decomposing leaves (they add humic , tannic, and carbonic acids to soil) and near the pH of 6.0. This is part of the cycle of decaying leaves returning to soil.

36. Depending on the specific nutritional needs of the species, the optimum pH can vary. This has a lot to do with how the species of plant has evolved over millennia to deal with its particular environment.

37. For instance, cranberries and blueberries, native to acidic peat bogs, prefer soil at around a pH of around 4.5. They will not grow well if soil is not mended with peat and humus.

38. Desert plants, like yuccas, on the other hand, may grow well at higher pH values, since the soil in many deserts is full of alkaline minerals and humus is scarce or absent.

39. For most plants, a pH value below 4.0 is harmful, because the acidic soil causes many essential nutrients to leach out of the soil. Aluminum salts and magnesium salts are also highly soluble in acidic soil, which is toxic to plants.

40. On the flipside, soil that is too alkaline is harmful, because it inhibits the solubility of many needed micronutrients like Fe, Co, Cu, and Mn, and the plant can't take them up to use as **coenzymes.**

41. In agricultural farming, all of these factors have to be taken together and balanced. The chart below shows the nutrient availability across a pH range, for several of the most common nutrients needed by plants.

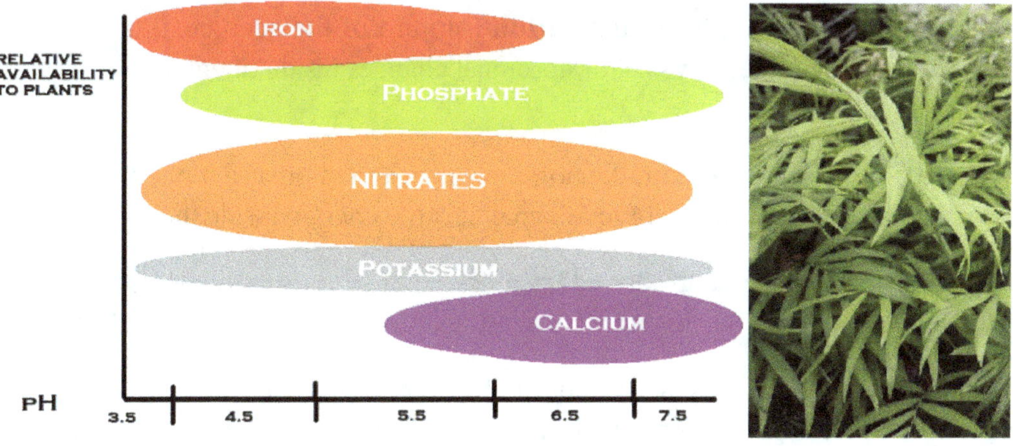

42. As you can see, iron becomes more soluble toward the acidic side of the scale, while calcium becomes more soluble toward the basic side of the scale. Therefore, a balance at 6.0 is necessary for most plants.

43. Because of the complexity of soil chemistry, over-fertilizing a field will usually do more harm than good.

44. The first problem comes in when a plant can't get enough OTHER nutrients to keep up with the limiting nutrient that it suddenly has in abundance.

45. It's no good having tons of phosphates to make ATP from photosynthesis, if the plant doesn't have any magnesium to build chlorophyll or iron to build cytochromes. The extra nutrient just sits there.

46. Additionally, over-fertilized crops undergoing a growth explosion may deplete the soil of nitrogen. It doesn't matter if other minerals are there if nitrogen is low. The plot won't be productive in later years.

47. This is why crop-rotation is practiced on many subsistence farms. Some plants return an excess of nitrogen to the soil (such as beans), while others use it up (like corn). However, this is not true for other minerals.

48. By rotating the crops strategically, you can complement nutrient use with supplementation from year-to-year.

49. Another more basic problem arises if excess fertilizer remains in the soil without being taken up. Because of **osmosis**, the extra mineral salts in the soil then begin pulling water out of plant roots.

50. Eventually, **salinization** occurs and the **water potential** of the soil becomes so low that crops can't grow until the excess nutrients are leached out of the soil by rainfall AND a new layer of topsoil is added.

51. While this section has very briefly discussed the needs for various nutrients for plant growth, understand that entire scientific textbooks can be written on the numerous effects single nutrients.

52. Soil chemistry is a hugely complex topic that is beyond the scope of this text, but interesting nonetheless. For further reading, look into agricultural and botanical journals or texts.

C) Photic and Thermal Growth Responses

1.Light plays a tremendous central role in the growth of plants, since photosynthesis produces the food that plants use for growth and metabolism. Additionally, its presence or absence triggers secondary responses.

2. The first place in a plant's life that **solar energy** levels triggers cellular signals occurs during seed **germination.**

3. Along with oxygen levels and water availability, **scarification** (rupture of the seed coat), the relative amount of light available serves as a trigger mechanisms for hormonal changes inside of embryos, triggering development.

4. When a seed is dormant, genes for **abscisic acid** production remain switched on at intermediate to high levels. This suppresses germination and keeps the embryo asleep in its little cocoon inside the seed.

5. Incoming light triggers receptors that, in turn, send **secondary messenger signals** to the genes controlling the production of **abscisic acid.** These genes begin to turn off and the dormancy comes to an end.

6. As light becomes increasingly available, **abscisic acid** stops inhibiting root growth. Other genes for **gibberellin** and **cytokinins** production turn on. Their products trigger cell division and embryo development.

7. Light also triggers **auxin** production in the shoot, that causes the embryo to develop leaves.

8. Differential levels of numerous types of these hormone classes guide the development of the embryo into a seedling, and then into an adult plant.

9. Light remains an important factor in switching genes on and off throughout the life of an adult plant.

10. For instance, **phototropism** occurs because of differential **auxin** production on opposite sides of the plant. Cells divide and lengthen more quickly on the side away from the light, causing the plant to lean towards the light.

11. In **heliotropism**, certain herbaceous plants can change their position towards the sun as the day goes on, because various hormones trigger potassium release or uptake, causing osmosis to reposition the stem.

12. Plants also **bloom, germinate,** and regulate their growth rates in response to seasonal light levels.

13. Many genetic cues are triggered by a specific number of hours of sunlight, alternating with a specific number of hours in the dark. This is why daffodils bloom on cue in February, while peppers will only bloom in Summer.

14. Different species of plants have different **photoperiod** requirements, mostly dependent upon the location where their species evolved. A **photoperiod** is a distinct period of time where there is light or darkness.

15. The table below describes the four types of **photoperiods** exhibited in the plant kingdom.

Photoperiod Type	Description	Examples	Picture
Short Day Plants	Flowering occurs when there are more hours of darkness than daylight. Typically, these are tropical species that bloom during the milder seasons of the year (but not always).	Poinsettia Chrysanthemum Christmas cactus Rice Soybeans Violets Some Onions Garbage Tree	
Long Day Plants	Flowering occurs when there is more light than darkness. Often triggered by the Spring equinox.	Jalapeno pepper Spinach Black-eyed-Susan Blue-eyed Kathy Green-eyed Karen Cucumbers Coneflower Hibiscus Potatoes	
Intermediate Day Plants	Blooming and growth are maximal when the day and night are approximately equal (12 hours).	Coleus Sugar cane Most grasses Most onions Goochweed	
Day-Neutral Plants	Factors other than light determine blooming and growth patterns. Most tropical plants are day neutral.	Corn Sunflowers Orchids Strawberries Tomatoes Common Stinker	

16. For instance, most plants that are from Arctic regions are **long day plants**, since they must get all their growing done during the Summer, where the sun may only go down for a few hours, if at all.

17. Many tropical plants are **day neutral**, meaning that light is not as large of a determining factor in triggering growth cues as other stimuli. This is because there aren't very distinct seasons near the equator.

18. Other sub-tropical and tropical plants are **short day plants**, since they try to time growing delicate blooms during the Winter equinox, when the weather is milder.

19. Many temperate plants and sub-tropicals are **day-intermediate plants** that bloom primarily during Spring.

20. So how do plants know what time of day it is? It's not like peach trees wear Rolexes. They do it by monitoring light levels with photo-sensitive chemicals called **phytochromes.**

21. **Phytochromes** are blue-green colored pigments that absorb light most efficiently near the red end of the spectrum. **Porphyrin rings** make up the center of these molecules. These are anchored to protein domains.

22. In a way, they are similar to hemoglobin and cytochromes, in that the protein domains on their perimeter position and hold he aromatic rings in the correct orientation to absorb energy.

23. In this case, it's light that the molecules are absorbing, rather than oxygen molecules or electrons.

24. **Phytochromes** are assembled from two different parts. The porphyrin part is made by the chloroplast, while the protein component is encoded by nuclear DNA and made by ribosomes in the cytoplasm.

25. After assembly in the Golgi, the finished phytochromes are released back into the cytoplasm.

26. There are five types of these pigments (**phytochromes A, B, C, D,** and **E**). All five are sensitive to slightly different wavelengths in the red spectrum, because of variations in the chemical structure of the porphyrins.

27. When near red light (660 nm) strikes a **phytochrome**, the photons provide enough energy to cause electrons to jump and change the molecular conformation, moving the position of a double bond in the ring.

28. Near red light activates phytochromes during the day. In turn, other secondary messengers respond to the active phytochromes by switching on specific light-responsive genes.

29. The active form of a phytochrome is called **Phytochrome Red** or the abbreviation P_R.

30. The active P_R molecule binds to a protein in the nucleus called **phytochrome interacting factor**. When combined, the two proteins make a **transcription factor**.

31. The protein transcription factor complex then binds at the **promoter** regions of many genes for growth and/or flowering or sprouting and causes them to be turned on and **transcribed.**

32. At night, far red light (730 nm) 'resets' the molecule and causes it to bend back the other way. The double bond switches back to its original position, in response to lower energy photons.

33. These far red wavelengths occur around dusk and sunset, which effectively turns off the light-activated genes that were active during the sunny parts of the day.

34. The inactive form of the molecule is called **Phytochrome Far Red** or the abbreviation P_{FR}.

35. This is much like the way light bends retinal in animal eyes bends into cis and trans shapes.

36. During the day, sufficient photons are available for color daytime vision, while at night, lower energy photons cause vision to fade to mostly black-and-white.

37. The diagrams below show the molecular structures of phytochrome red and phytochrome far red.

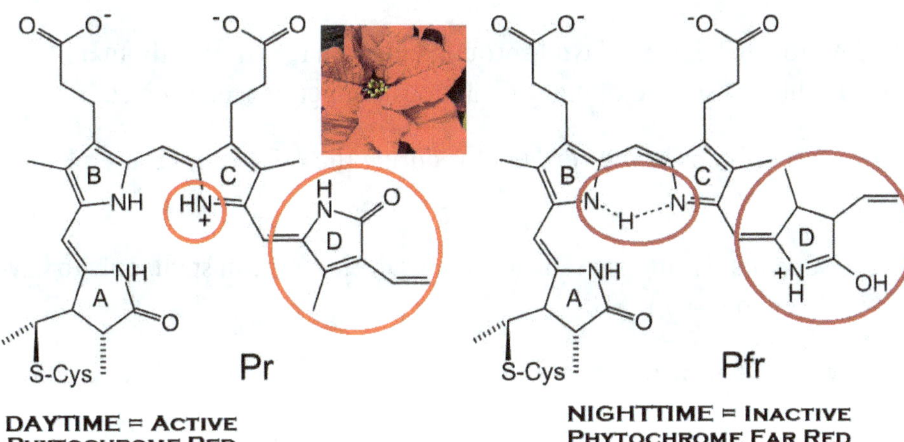

38. **Phytochromes** are so sensitive to red light, that even a single flash of red light in the middle of the dark part of a photoperiod will interrupt blooming or growth.

39. For instance, if you place a flashing red light bulb next to a Christmas cactus, it will fail to bloom because it's biological clock has been screwed up.

40. This also explains why most houseplants never bloom. They just don't get the correct ratio of dark to light, so the genes that control flowering are never turned on.

41. The diagram below summarizes some of the effects that light and temperature have on various plants.

42 Many people get poinsettias for Christmas and keep them as houseplants, unable to get them to bloom. This can be changed if they are put into a dark closet for the correct number of hours each day for several weeks.

43. One famous set of experiments used cocklebur plants (a weed that is a known long-day plant) and interrupted their daily cycles with flashes of red light. Even a brief flash in the dark period will stop them from blooming.

44. In addition to light, **temperature** also plays a role in triggering certain growth events in a number of different plant species.

45. Many plants, such as potatoes and carrots will have delayed sprouting (or won't sprout) without a period of cold dormancy called **vernalization**. This is also true of many deciduous temperate tree species.

46. This is also why many people store flower bulbs in the refrigerator during the offseason before planting them.

47. This is a protective mechanism against plants putting out delicate shoots during a warm spell, only to have them frozen off a week or two later by a later cold snap.

48. In most plant species, blooming flowers and growth of new shoots in flowers in the Spring is a combined effect of **photoperiod** and **vernalization**.

49. Happy trails. Go make yourself some dessert. You've earned it, you supah-star! 8 BALL!

'Eight Ball' Monocot Cake Recipe

In celebration of monocots, this cake contains ingredients from EIGHT different monocot plant species. This will CERTAINLY be a conversation piece at your next social event, opening the door for such topics as the pros-and-cons of C3 vs C4 metabolism, the definition of true wood, and the shocking revelation that what you thought were petals....well they might just be SEPALS!

CAKE

2 1/2 cups of cake flour MONOCOT #1 is WHEAT
1/2 cup of rice flour MONOCOT #2 is RICE
1/2 teaspoon of baking soda
1 teaspoon of salt
1 cup unsalted butter
2 cups granulated sugar MONOCOT #3 is SUGAR CANE

6 large eggs (take out of fridge 30 min)
2/3 cup of sour cream (take out of fridge 30 min)
3 teaspoons of vanilla extract MONOCOT #4 is VANILLA
2 teaspoons of coconut extract MONOCOT #5 is COCONUT
1/2 cup of coconut milk
1/2 cup of cream of coconut (NOT the same thing!)
1 cup of shredded fresh coconut
1 teaspoon of corn syrup MONOCOT #6 is CORN

COOKING

Preheat oven to 350. Whisk flour, sugar, baking powder, salt together and set aside. Turn on electric mixer and dump in butter and sugar. Let it run 2 minutes. After they are combined, dump in your sour cream, cream of coconut, corn syrup, and coconut milk. Grease 4 cake pans (9 inch) and use a spatula to get all the batter into pans. Bake for 22 minutes. While these are going, make your icing.

ICING /TOPPING

1 stick of salted butter
1 brick of cream cheese
6 cups of powdered sugar
1/4 cup coconut milk

1 teaspoon of vanilla extract
1 teaspoon of coconut extract
1/4 teaspoon of salt
2 cups of fresh shredded coconut

Handful of chopped candied dates MONOCOT #7 is DATES
Handful of chopped candied pineapple MONOCOT #8 is PINEAPPLE

Let the butter and cream cheese sit out while the cake is baking to soften. Use a hand mixer to whip them together until homogenized. Add remaining ingredients OTHER THAN COCONUT and keep mixing until even. Add sugar or coconut milk to adjust texture. WAIT AT LEAST AN HOUR after the cake has cooled to ice it, or you will be sorry! It will melt and people will laugh at your cake. Sprinkle the coconut into the sticky icing. Chop up some candied dates and pineapple and toss them over the finished cake carelessly. 8 BALL!

NOTE

You can add candied ginger, rolled oats, barley, lemon grass, tequila, roasted garlic, or even grass clippings from the yard to up your monocot count!

CAKE TRIVIA
Answer page **192**

1) Why is a pound cake <u>called</u> a 'pound cake' anyway?
2) What type of cake did Billy the Kid request of Pendleton in his confrontation with Mr. Chisum in Young Guns II?
3) Who was incorrectly credited with the inflammatory phrase 'Let them eat cake...' in the French Revolution?
4) What ingredients (though you probably shouldn't) CAN be substituted for oil and eggs, respectively, in some cakes?
5) What is the secret ingredient in German Black Forest cake?

Chapter Eight

GLOSSARY

ABSCISIC ACID: A plant hormone derived from carotenoids that slows growth and counteracts the effects of auxins and other growth promoters. Manufactured in response to environmental stressors. Promotes seed dormancy and dropping of leaves.

ACCESSORY FRUIT: A fruit that is derived from non-reproductive stems or other components of the plant other than the ovary wall. Strawberries and figs are examples. The true fruits are usually dry fruits that surround the seed in these false fruits.

ACHENE: A small dry fruit that encloses a single seed. There are no natural seams on the ectocarp, meaning that it has to be splintered to open. Sunflower seeds, quinoa, the true fruit (casings) on strawberry seeds and caraway seeds are examples.

ACORALES: Order of monocots that includes some of the most primitive members of the group. Members have blade-like leaves that produce oily secretions, tiny flowers emerging from a spadix, and monosulcate pollen. Commonly called sweet flags.

ACTIVE TRANSPORT: Process used by cells to forcibly take substances in against their concentration gradient. Requires the expenditure of ATP energy on part of the cell. Includes use of pump proteins (antiports, symports, uniports) and endocytosis.

ADAPTIVE RADIATION: Evolutionary process, wherein a single species adapts and evolves to fill numerous open ecological niches, reducing competition for resources, and rapidly diverging into multiple different species over a relatively short period of time.

ADHESION: The ability of water molecules to stick to other slightly charged molecular surfaces and create surface tension. Caused by the dipole moment of water, where the slightly negatively charged oxygen atoms adhere to the slightly positively charged hydrogen atoms of neighboring molecules, such as the cellulose xylem in a tree trunk.

ADNATE ANTHER: Found in primitive basal dicots like magnolias, it is a type of anther where the stamens attach to the filament along the entire length of their dorsal surfaces.

AGGREGATE FRUIT: A type of fruit that is really multiple ovaries fused together that emerged as multiple pistils from the same flower stalk. Each fruit contains its own seeds. Examples include raspberries, blackberries, pineapples, and strawberries.

ALISMOTALES: A primitive order of monocots that retain certain algal features, such as chlorophyll in the embryos and an aquatic germination stage. Their seeds lack endosperms and they have monosulcate pollen. Examples include arums and sea grasses.

ALKALOID: A class of (usually) toxic compounds derived from nitrogen rich amino acids, such as tryptophan, that plants manufacture to deter consumption by herbivores. Particularly common among members of the nightshade family. Examples include atropine, nicotine, strychnine, ephidrine, and caffeine.

ALTERNATION OF GENERATIONS: System of reproduction evolved in green algae and retained by plants (with great modification in most cases), in which an asexual spore-producing haploid adult life sporophyte stage alternates with a sexual gamete producing diploid or polyploid gametophyte stage. Originally drought favored the sporophyte, while rainfall favored gametophyte growth.

AMINO ACIDS: Class of biomolecules that are the monomers (building blocks) of proteins. There are 20 biologically important amino acids in plants. They are manufactured by various biochemical pathways that represent the convergence of Calvin cycle derivatives and nitrate uptake from the soil. Needed to form proteins, alkaloids, hormones, and other secondary compounds.

AMMONIFICATION: Biochemical pathway used by various soil bacteria involved in the decomposition of proteins and nucleic acids from dead tissues and waste. The bacteria use the carbon skeletons, while the ammonia molecule is cleaved off and used by nitrification bacteria, in turn making nitrates available to plants.

AMYLOSE: Polysaccharide commonly known as starch, which is the primary means of long-term carbon storage in almost all plants. Stored in roots, stems, and sometimes leaves, it is a long polysaccharide polymer of alpha-linked glucose monomers.

ANGIOSPERM: The dominant clade of plants in existence today, it comprises all flowering plants that produce seeds inside of fruits. Other commonalities include larger numbers of vessel elements in their vascular systems, endosperm in the seeds, microscopic sporophyte-dependent gametophytes inside flower structures, and use of symbiotic animals for pollination and seed distribution.

ANNUAL: An herbaceous plant that uses an extreme R-strategist lifestyle, wherein it grows rapidly in a single season before shifting its nutrient reserves to production of numerous flowers and seeds, completing its entire life cycle in a single season. Common in species found in temperate and colder latitudes with short growing seasons. Examples are marigolds, zinnias, sunflowers, cosmos.

ANTHER: In angiosperms, the organs that serve an analagous purpose as testes in male animals. Their germ cells divide by mitosis and then microspore mother cells divide by meiosis to produce pollen grains. Situated atop the filament, they comprise the other part of the stamen. In perfect flowers, they are found in a whorl outside the pistil.

ANTHERIDIUM: The male gametophyte in bryophytes and pterophytes that produces primitive sperm by meiotic division. These sperm rely on droplets of water to swim to the ova inside the female archegonium. Lives as separate plant from female gametophyte plant in bryophyt and many pterophytes . Occupies a separate region of tissue on the gametophyte of ferns.

ANTHOCYANIN: Water-soluble plant pigments derived from amino acids and sugars that are various shades of red, blue, and purple, depending on pH and sidechain chemistry. Absorb many of the green and yellow wavelengths that other pigments cannot. Examples include the purple coloration of grapes and eggplants, blue hydrangea flowers, and red roses and apple skins.

ANTIPODAL CELLS: Accessory cells found in plant embryos that have the responsiblity of nourishing and caring for the developing embryo. In monocots, they impregnate the endosperm with lipids, starch, and proteins. In dicots, they pack the cotyledons.

ANTIPORT: A type of membrane protein used in active transport that forces one substance out against its concentration gradient, while exchanging it for another substance that is forced in. Requires ATP energy. Example is the sodium-potassium pump.

APIALES: An order of dicot plants characterized by bottle-shaped flower pistils, bifurcated ovaries, and aromatic compounds in the sap. Commonly have annual or biennial life cycles, but some members are also woody perennials. Includes carrots, parsley, mock orange, dill, gensing, cumin, and tropical scheffeleras.

APICAL MERISTEM: An area of growth that emerges from undifferentiated tissue at the shoot tip and root tip of a plant. Forms primary root and primary stem growth responsible for adding to the height of a plant and depth of roots.

APOPLAST: The space outside of a cell membrane where fluids and dissolved substances travels freely in empty space. In plants, this space is bounded by the cell wall. Therefore, plasmodesmata are needed to join apoplasts of neighboring cells.

AQUAPORINS: Tiny donut-shaped proteins impregnated in all cell membranes that act as tunnels to allow the passive transport of water into and out of cells. In osmosis, these tunnels allow water to move, but solutes are too large and are blocked from entering these. Therefore, they must go through their own carrier proteins to gain access.

ARCHEGONIUM: The female gametophyte of non-flowering bryophytes and pterophytes. Germ cells in the archegonium produce ova from their germ cells by meiotic division. Eventually these ova develop into spores after they are fertilized by free-swimming sperm. In bryophytes and many pterophytes, these are separate plants from the male, while they are areas of tissue in the hermaphroditic gametophytes of ferns.

ARECALES: Order of monocots that includes palms, such as coconuts, dates, acai, oil, saw palmetto, and hundreds of other species. Characteristics include false trunks impregnated with stiff fibrous ground tissue, growth of whorl of fronds or fans starting at apical meristem crown, spadix flowers, absence of starch in seeds, drupe-like fruits, C4 metabolism.

ASEXUAL REPRODUCTION: System of reproduction mediated by mitosis, wherein clones of the parent are created. There is a tradeoff of rapid reproduction and minimal resource investment in offspring in exchange for lack of genetic diversity. In plants, budding and fragementation are both common in many orders.

ASPARAGALES: The largest order of monocots comprising nearly 300,000 species including asparagus, orchids, aloes, irises, agaves, onions, daffodils, and many others. Often host root mycorrhizae essential to their robustness. Typically are herbaceous and sword-like paralllel leaves grow in a rosette from apical meristem. Flowers usually have tepals in parts of six.

ASTERALES: Very large order of dicot plants that includes asters, daisies, zinnias, sunflowers, bellflowers, and elderberries. Common features include hollow pollen-filled tubular stamens, carbohydrate storage of inulin instead of starch, flower parts in fives.

ATP: Adenosine triphosphate, the energy currency of cells. All biochemical compounds produced as stored foods in plants are ultimately converted to ATP through cellular respiration. Almost all processes in cells that require work, such as transport of materials, require ATP to do so. Also produced during the light reactions of photosynthesis.

AUXINS: Class of plant hormones that have numerous effects on growth, depending on which type of auxin is involved. Derived from tryptophan and similar amino acids, their side chains determine which receptors are triggered. Effects include cell division and growth at the apical meristems, lengthening stems and roots, but also rooting, fruiting, and inhibition of side branches.

AXILLARY PLACENTATION: Found in fruits where the hilum of the seeds attaches to a central placenta derived from the ovary wall. Seen in peppers, tomatoes, and citrus fruits.

BARK: Protective outer covering over woody plants. Composed of all tissues outside the living vascular cambium. Layers include dead cork cells, cork cambium, and dead secondary phloem, as well as fibrous sclerenchyma and collenchyma cells.

BASAL MAGNOLID DICOT: The most primitive group of dicots that emerged as some of the first flowering plants. As compared to more advanced dicots, they have more tracheids in their vascular tissue, simple monosulcate pollen, crude flowers, and simpler ovarian and vascular bundle anatomy. Includes magnolias, water lilies, anise, and many aquatic weeds.

BASAL PLACENTATION: Found in achenes like sunflowers, the seed attaches at the base of the ovary and takes up a large portion of the internal space of the ovary.

BERBERIDOPSALES: An order of relatively primitive dicots that share the common features of scaly trunks and stems and opposite leaf arrangement. Besides these features and a few common flower similarities, this group is largely classified by molecular evidence. Includes mistletoe, sandalwood, witches broom, coral plant, and a number of desert xerophyte shrubs.

BERRY: A type of fruit that grows from a single ovary that has multiple ovules. Typically, the ectocarp is fleshy, the mesocarp is soft (and often contains sugars and attractant compounds), and there is a soft endocarp around the seeds. True berries include blueberries, cranberries, grapes, and pokeweed. Pepos and hesperidiums are also larger and modified types of berries.

BICARPELLARY FLOWER: A flower that has two pistils at the center of the corolla. This is independent of the number of locules in each ovary. Found in many types of plants, including brassicas.

BIENNIAL: A type of plant with a two-year life strategy. In the first year, the plant grows lots of foliage and stores back starches, usually in a large taproot. In the second year, the plant mobilizes these starches into the production of numerous seeds and reaches the end of its lifespan. Used mostly by plants, such as carrots, celery, and parsley in temperate and cold climates.

BILOCULAR OVARY: An ovary that uses the ovarian wall to form two distinct divided lobes with separate seed compartments. A common type of arrangement in members of the nightshade family of in Order Solanum.

BLADE: The flat photosynthetic surface of a leaf. The blade is comprised of layers of chloroplast-containing mesophyll tissues and vascular bundles sandwiched between an upper and lower epidermis.

BORAGINALES: An order of dicots that usually share the features lance-shaped leaves, production of toxic alkaloids and secondary compounds, and the presence of leaf trichomes. Includes the buglosses, boIrage, comfrey, and heliotrope.

BRACT: Modified leaves or scales that protect a developing flower cluster and/or help to position them away from the stem for pollinators. Some bracts appear to be petals, but take the place of petals to attract pollinators, as in dogwoods and poinsettias.

BRASSICALES: Large order of dicots usually (but now always) native to cooler and temperate climates. Includes cabbages, kale, bok choi, mustard, nasturtium, capers, and papayas. Common features include racemose flower stalks, nectaries between petals and stamens, production of sulfurous oils to deter insects, and use of brassicosteroids to regulate their growth.

BRASSINOSTEROIDS: Class of plant hormones first discovered in cabbage family plants, but appear to be present in all plant tissues. Derived fromCcholesterol. Responsible for wide-ranging effects, including stimulation of xylem growth, cell division, stem elongation, root inhibition, development of flowers, and numerous metabolic changes according to environmental cues.

BRYOPHYTA: Class of primitive non-flowering spore plants that includes mosses, hornworts, and liverworts. They lack true vascular tissues, pollen, and flowers. They are diecious during the gametophyte stage, with separate male and female plants. Their leaves and stems are primitive and they have rhizoids instead of true roots. Have stomata in sporophyte stage.

BUD SCALE SCAR: Found in woody stems, these scars create concentric rings around the circumfrence, marking the location of the apical meristem buds in prior years, giving a relative indicator of how much growth occurred each year.

BUDDING: Process of asexual reproduction where a parent produces identical miniature clones of itself by mitotic division of undifferentiated tissues. In plants, meristems divide to create daughter clones. Examples include plantlets in mother-of-thousands kalanchoe, runners in strawberries and kudzu, pup plants in crepe myrtles, and shoots in bamboo.

BULB: A cluster of non-photosynthetic underground starch-filled leaves growing in concentric layers, such as in onions. These specialized leaves take the place of the typical storage functions of the root or stem. Examples include garlic, onions, and daffodils.

BUNDLE SCAR: Cluster of dot-like impressions found in the center of leaf scars, indicating the prior location of vascular bundles in that leaf before the petiole separated from the stem during abscision. Seen in deciduous trees.

BUNDLE SHEATH CELLS: Cluster of specialized mesophyll cells (ground tissue) that surround the xylem and phloem vessels in the leaf, much like plastic around an electrical cord. In all plants, they regulate the transport of sugars into the vessels. In C4 plants, they perform several steps of photosynthesis, storing malic acid, which is split to release carbon dioxide inside of the plant's tissues. This allows the plant to continue running photosynthesis with the stomata closed, preventing water loss.

C3 METABOLISM: Photosynthetic pathway, so called, because the Calvin cycle produces the 3-carbon end product, phosphoglyceric acid (PGA), which is often doubled-up and linked to produce glucose. The cost of C3 metabolism is 18 ATP and 12 NADPH from the light reactions, with a return of glucose molecules that produce 36 ATP when split. C3 requires open stomata to bring in CO2, which is a disadvantage during hot, dry times.

C4 METABOLISM: Photosynthetic pathway, so called, because PEP-carboxylase enzyme first takes up CO2 and transports it to the bundle sheath cells, which create the 4-carbon malic acid. While this comes with extra cost of 12 ATP in pumping, it allows tropical and desert plants to continue photosynthesizing with closed stomata in hot, dry conditions, as CO2 is cleaved off the malic acid and used in the regular Calvin cycle.

CALVIN-BENSON CYCLE: Autotrophic biochemical in plants that manufactures phosphoglyceric acid (PGA) from 18 ATP and 12 NADPH in the light reactions, by essentially joining CO2 molecules onto cycling carbon skeletons by using the energy from the ATP and they hydrogen atoms transported by NADPH. In the end, the PGA can be sent through multiple biochemical pathways, but most frequently forms glucose or fructose.

CALYPTRA: Found in the sporophyte of mosses, it is a protective fibrous case that surrounds germ tissue as it divides by meiosis into haploid spore. Once the spores are mature, it dries and breaks open, scattering spores.

CALYX: In angiosperms, the whorl of sepals outside the petals that surround and protect a developing bud.

CAM METABOLISM: Used by a few taxa of tropical plants and xerophytes like cacti, it creates malic acid like C4 plants, but does not actively pump malic acid into bundle sheath cells to prevent photorespiration. Instead, they make their malic acid in the mesophyll by opening stomata only at night. CO2 can then be released to run the regular Calvin cycle the following day with teh stomata closed.

CAMPANULID FLOWER: A sub-clade of Euasterid dicots that are characterized by disc-shaped germ-line cells. Star-shaped petals are derived from the outer perimeter, while pistils and stamens emerge from cells in the middle.

CAPILLARY ACTION: The combined action of adhesion and cohesion that causes water to rise upward over narrow spaces with a force stronger than gravity. This allows water to travel up the xylem of large trees to the leaves.

CAPSULE: A type of dry dehiscent fruit that results from the fusion of multiple ovaries into one structure. Typically, there are many ovules that develop into a large number of seeds. Examples include okra, cotton, poppies, and lilies.

CARBOHYDRATES: Class of biochemical compounds that consist of carbon rings with a 1:2:1 ratio of C:H:O. Monosaccharides, such as glucose and fructose are primarily used for energy and for nectar. Monosaccharide monomers can be chained together to form polysaccharides such as starch, cellulose, and pectin.

CARNIVOROUS PLANT: Plants that use various trap mechanisms to lure insects and other small animals into a digestive enzymatic chamber to extract nitrogenous molecules from their tissues. This strategy is found in plants, such as pitcher plants and Venus fly trap, native to areas with anoxic swamp mud with denitrification bacteria.

CAROTENOID: Lipid-soluble orange-colored pigments, especially Beta-carotene, that are derived from isoprene and sterols. Plants use them to absorb high energy purple, blue, and dark green wavelengths for photosynthesis.

CARYOPHYLALLES: A diverse order of asterid dicots that share common biochemical markers, often have CAM or C4 metabolism with PEP carboxylase carbon fixation, large bundle sheath cells in the leaves, and adaptations to survive drought and heat. Includes cacti, amaranths, marigolds, portulaca, rhubarb, and many different succulents.

CARYOPSIS: A type of dry indehiscent fruit with a fused fruit endocarp and seed coat that are indistinguishable. Many members of the Order Poacea produce this type of seed, including wheat, barley, oats, corn, and rice.

CASPARIAN STRIP: A waxy barrier in the root endodermis formed by thickened cell walls on the exterior surfaces of cells that prevent backflow of water from the vascular system in the interior root back out into the soil.

CELL WALL: Composed of primarily cellulose and lignin, it is a mesh-like protective fibrous structure located outside of the cell membrane. It is usually composed of multiple layers glued together with pectin. Provides structural support and prevents the osmotic rupture of plant cells in a hypotonic solution.

CELLULOSE: Complex carbohydrate composed of glucose monomers alternating in a zig-zag of 1,4 alpha and beta linkages. Additionally, 1,6 linkages with other chains of cellulose allow the formation of the fibers found in cell walls that create structures, such as fibers, seed coats, and wood.

CHEMOTROPISM: Tendency of a plant to grow toward a chemical stimulus. Occurs because molecules trigger receptors in plant cells that send secondary messages inside the cells, resulting in production of auxins and other hormones that cause cell division and growth toward that side.

CHLOROPHYLL A: Used in photosystem II of photosynthesis, it is a lime-green pigment found in the reaction center of thylakoids. When struck by light photons, excited electrons jump off the pigment molecule to a series of electron carriers that create an electrochemical gradient, getting protons to chase the energized electrons.

CHLOROPHYLL B: Used in photosystem I of photosynthesis, it is a khaki-green pigment found in the reaction center of thylakoids. It receives electrons from photosystem II, which are once again struck and energized by light photons and sent down an electron transport chain that produces 18 ATP and 12 NADPH when complete

CHLOROPHYTA: Phylum of eukaryotes that includes green algae. While single-celled, colonial, and filamentous algae grow in forms different than plants, many macroalgae resemble them. The putative ancestor of plants, they also have chloroplasts with thylakoids, store their food as starch, and have cellulose cell walls.

CHLOROPLAST: Organelle responsible for the manufacturing of biochemical food compounds in autotrophic eukaryotes such as green algae and plants. Consists of stacks of thylakoid solar panels full of chlorophyll, where light reactions are run, and an enzymatic soup known as the stroma, which runs the Calvin cycle and completes the biochemical pathways that make sugars and other molecules. All of this is wrapped up by an outer membrane.

CLIMAX COMMUNITY: In ecology, it is the penultimate mature community of organisms found in an ecosystem, especially forests. Typically, the base of the food web in climax communities is formed by old growth mature hardwood dicot trees that provide food in the form of fruit, nuts, or foliage.

CLONE: In botany, it is an exact genetic copy of a parent organism formed by mitotic asexual reproduction via budding or fragmentation. Pups, runners, leaflets, and shoots can all be parental clones.

COENYZMES: Organic molecules that bind at certain allosteric sites on enzymes, assisting them in their functions and helping to regulate reaction rates.

COEVOLUTION: A phenomenon that creates mutualistic symbiosis over time as two or more species evolve and adapt together to the point of dependency. This is true for numerous flower pollinator relationships, nitrogen-fixing bacteria and legumes, and for trees and mycorrhizal fungi.

COFACTORS: Typically, metal ions that interact with various amino acid residues in the protein chains of enzymes and pigments. They are critical to the proper function of these proteins and the primary reason why plants need metal ions found in soil.

COHESION: The phenomenon caused by the consequence of hydrogen bonding, wherein neighboring water molecules stick together because of attractions between opposite ends of their dipole moments. The slightly positively charged hydrogen atoms stick to the slightly negative oxygen atoms of their neighbors, forming a lattice and a rising water column over a narrow space, such as inside the xylem.

COLEOPTILE: In monocots, a protective sheath that surrounds the epicotyl of germinating shoots.

COLLENCHYMA: A sub-type of ground tissue that has several layers of cell walls. It forms stringy, flexible, strong fibers that support herbaceous stems and forms reinforcements in leaf tissues and fruit. Examples include the strings in celery stalks, the fibrous mesh in tough leaves like yucca, and the stalks of okra.

COMMELINALES: An order of mostly tropical monocot plants that includes water hyacinths, spiderworts, and dayflowers. Their leaves emerge in a whorl from a meristem protected by a sheath. They have both hermaphroditic and additional male flowers, and they only offer edible pollen to insects, but no nectar.

COMPANION CELL: In the phloem, a living cell that nourishes and cares for the helpless adjacent hollow seive tube cells that lack their own nuclei and exist solely to pump sugars through the plant.

COMPOUND FLOWER: A flower that is really a cluster of many different pistils inside a shared calyx of common petals and sepals. They produce multiple independent fruits when they go to seed. Examples include sunflowers, zinnias, marigolds, roses, mums, dahlias, and hundreds of other common ornamentals.

COMPOUND LEAF: An arrangement where a cluster of multiple leaves emerge from a common petiole. Pinnate or palmate compound leaves emerge from a common center petiole, while twice-pinnately compound leaves have leaves on smaller petioles that emerge from a larger center petiole.

CONE: Reproductive structures found in gymnosperms. Male cones are derived from clusters of leaves at the tips of branches and house microspore mother cells that produce pollen by meiosis. Female cones are derived from modified branches and house megaspore mother cells that produce ova by meiosis. While cones protect developing seeds, they are naked and provide no fruit to entice animals to distribute the seeds.

CONIFEROPHYTA: Phylum of cone-bearing seed plants that includes true conifers, cycads, gnetophytes, and gingkos. Common characteristics include naked seeds that develop within the protection of female cones, pollen-producing male cones, wood fiber that is mostly tracheids, needle or scale-like leaves (in most cases), and secondary compounds to prevent insect invasion.

CONVERGENT EVOLUTION: An evolutionary pattern where unrelated or distantly related organisms evolve similar morphology as a result of common environmental pressures. An example is spines and succulence in cacti and euphorbias, which in spite of near-identical appearances, in some cases, are not closely related at all.

CORK CAMBIUM: A living lateral meristematic tissue found in woody plants between the bark on the outside and the cortex (if any remains) and primary phloem on the inside. It divides outward to produce layers of cork in bark.

CORM: An underground stem filled with starch and stored nutrients, such as in irises, gladiolus, and crocuses. In addition to fibrous stems, tehy can be distinguished from bulbs by a basal plate where roots emerge, and modified scale-like leaves that form a tough protective cover.

CORNALES: An order of plants in the Asterid clade that are usually short trees and woody shrubs that produce drupe fruits from small clusters of numerous crude flowers surrounded by single bracts that look like petals. Many members are deciduous and produce anthocyanins in their leaves and fruit.

COROLLA: Refers, collectively, to all the petals that surround the reproductive parts of a flower.

CORTEX: A type of ground tissue found in stems and roots of herbaceous plants and saplings that is composed of rounded cells full of food storage vacuoles and leucoplasts.

COTYLEDON: The first embryonic leaf (monocots) or leaves (dicots) to emerge. Instead of having a photosynthetic function they are filled with starch and other nutrient supplements to help the embryo get a head start.

CUTICLE: A protective waxy layer secreted exterior to the upper and lower leaf epidermis that prevents water loss.

CYTOKININS: A class of plant hormones that encourages cell division for growth. They serve an important role in the proliferation of cells in leaves, encouraging chloroplast duplication, and preventing starch storage there.

DAY INTERMEDIATE PLANT: Refers to a type of plant that flowers in a hormonal response to days and nights of approximately equal lengths. Examples of such plants are artichokes, asparagus, onions, sugarcane, and many other monocots, such as grasses. Typically, flowering occurs in Spring or Fall.

DAY NEUTRAL PLANT: Plants that are not triggered to flower on the basis of day length. Their flowering is usually timed at specific points in their lifespan, independently of phytochrome control. Examples include members of the nightshade family, such as tomatoes and potatoes, cucurbits, and certain legumes.

DECIDUOUSNESS: The ability of certain dicot trees to shed their leaves in response to dropping levels of light and temperature, so that they can go into dormancy during winter, when it would be more energetically costly to photosynthesize than what the plant would return in stored food. May also be a response to dry seasons in tropical species. Examples include maples, oaks, hickories, and dogwoods.

DEHISCENT DRY FRUIT: A type of fruit where the ovary becomes a dry, fibrous protective coat, rather than a fleshy enticement for animals to feed and spread seed. Dihescent fruits have natural seems that split open to release seeds. Examples of such fruit include sweet peas, green beans, milkweed pods, and mustard seed pods.

DENITRIFICATION: In the nitrogen cycle, this step is controlled by bacteria that utilize nitrogen compounds such as nitrate and ammonia for energy, resulting in the removal of nitrogen gas from the soil back to the atmosphere. Plants in these oxygen-poor soils must adapt to take up enough nitrogen.

DERMAL TISSUE: One of the three basic types of plant tissues, it forms a protective barrier between the ground tissues and vascular tissues of the plant and the environment. Among its functions are sealing of leaves and stems to prevent water loss, uptake of CO_2 through stomatal openings, and absorption of materials through roots.

DICOT/DICOTYLEDON: One of two major divisions of angiosperms, dicots are characterized by networked leaf veins, a dartboard pattern of vascular bundles in the stem, xylem at the center of the root, a pith and cortex in herbaceous stems, an exterior cortex in roots, and sometimes the ability to form woody tissues as secondary growth. Their flowers have parts in multiples of four and five and their seeds have two cotyledeons.

DIDYMOUS STAMENS: Found in gardenias, jasmine, and many other plants among several orders, the stamens all sit in a whorl at an equally elevated height above the flower.

DIDYNAMOUS STAMENS: Common among members of Order Lamiales, such as olives and mints, this is a tetrad of stamens, wherein one pair of central stamens rises above the level of two inferior exterior stamens.

DIFFERENTIATED TISSUES: These are tissues with definite function and histology. Derived from meristematic tissues, they have been genetically activated to become specific types of cells with specific jobs, such as palisade mesophyll in leaves or phloem in the stem. In many plants, these tissues cannot produce clones of themselves.

DIFFUSION: The physical process, wherein particles even out their concentrations in a medium over time, with particles in areas of high concentration traveling to lower concentrations, due to the law of entropy. In plants, the uptake of CO_2 is governed by diffusion.

DILLENIALES: A relatively ancestral order of dicot angiosperms that retain primitive magnolia-like flowers lacking sepals, deeply defined secondary leaf veins, and DNA sequences that imply that they diverged from most other dicots long ago. Examples include peonies and elephant apple.

DIOECIOUS: A type of plant that has separate male and female individuals, rather than employing monoecious hermaphroditism. Many primitive plants, such as bryophytes and pterophytes (other than ferns) use this system, as do gingkos and cycads among the gymnosperms. Some angiosperms like muscadines, persimmons, and kiwi fruits also have separate genders.

DIPLOID: A cell that has two copies of each chromosomes in each cell nucleus, as they have one set of paternal chromosomes and another set of maternal chromosomes. Though common in plants, ploidy is not a given as it is in animals, where very few species are exceptions to diploidy.

DISPACALES: An order of dicots that are usually woody shrubs and vines that commonly have opposite leaf arrangements, glands in the leaves that secrete aromatic compounds, and trumpet-shaped flowers with fused petals. Examples include honeysuckle, weigela, snowberry, and viburnum.

DISTRACTILE ANTHERS: In pincushion flowers and other members of Order Dipsicales, the stamens face at perpendicular angles to one another. The larger stamen is vertical, while the smaller one is located further down the filament and faces horizontally.

DIVERGENT ANTHERS: In trumpet viens and other members of the Order Lamiales, the bases of each stamen are anchored to the filament, but their apexes point away from one another in opposite directions.

DIVERGENT EVOLUTION: An evolutionary process, wherein a closely related taxa of organisms becomes separated by geographic barriers (allopatry) or some other factor such as behavior or niche (sympatry), changing into different species under varied environmental influences. Often, anatomical features with much different morphologies are derived from the same embryonic structures, such as the paw of a cat and the wings of a bat.

DORSIFIXED ANTHERS: Found in members of the Order Malpighiales, these anthers attach to the filaments from the middle of the dorsal surface to the bottom end of the dorsal surface.

DRUPE: A type of fleshy fruit with a hard stone-like single seed with a fibrous pericarp loaded with very hard sclereids and cells with multiple cell walls. Examples include cherries, peaches, plums,walnuts, and almonds.

ECOLOGICAL NICHE: All of the specific ecological roles and relationships attributed to a species in a given habitat. For example, bananas are understory vegetation that live on the edge of rainforests and provide food for many herbivores and omnivores, such as hornbills and monkeys.

EMBRYO: The developing rudimentary stage of a plant that is developing early leaves (epicotyl), stems (hypocotyl), roots (radicle), and food stores (cotyledon(s) and endosperm. It develops from spores in bryophytes and pterophytes, but inside a protective seed in gymnosperms and angiosperms.

ENDOCARP: The portion of a fruit that surrounds the seed, often with a protective function. Examples include the fibrous core of an apple, as well as the outside stone layer surrounding a peach pit or almond.

ENDOSPERM: A nutritive component of a seed that is filled with starch, lipids, and proteins for the developing embryo. It has an important independent function in monocot seeds, but becomes incorporated into the cotyledeons in dicots. It is triploid, developing from a diploid germ cell fertilized by the second sperm inside of a pollen grain during germination.

EPHEDRINE: A protective alkaloid by members of the ephedraceae family of gnetophyte plants. It has a potentially toxic stimulant effect on the heart and central nervous system of animals.

EPICOTYL: The region of developing cells in an embryo that are fated to divide to become the apical meristem and give rise to the first true photosynthetic leaves. It is below the cotyledons.

EPIDERMIS: The outermost layer of the upper and lower surface of leaves and the outer layers of herbaceous green stems and roots. In leaves, they secrete a waxy cuticle and help prevent water loss. In stems and leaves, the stomata are located in this layer. In roots, it forms the outer surface and root hairs involved in absorption.

EPIGYNOUS PISTIL: Refers to flowers that protectively tuck must of their pistil away inside the pedicel. None of the ovary is exposed and only the tip of the stigma emerges past the corolla to allow for pollination.

ERICALES: An order of dicots that are colloquially called heathers. They have symmetrical flowers in multiples of five with fused petals and usually produce dry dehiscent capsule fruits. They are typically woody shrubs or ground covers. Examples include heathers, heaths, phlox, and primroses.

ETHYLENE: A simple alkene gas that acts as a ng fruit to pripengrow and , flowering to go through a developmental scycle, and leaves to shed, so that the plant puts more energy into reproducing. Auxins stimulate its production, while red light inhibits it.

EUASTERID CLADE: A clade of relatively advanced dicots that includes 15 orders of plants. In addition to typical dicot seed, leaf, stem, and root anatomy, they have petals fused to each other and to the stamens at the base of the flower, and they have a single integument around the seeds.

EUDICOT: The most advanced clade of angiosperm dicots, they are characterized by all of the classic features common to all dicots, plus the presence of three openings on their pollen grains and flowers that bear no resemblance to cones, as they do in lower dicots, such as magnolias.

FABALES: The order of angiosperm dicot plants commonly called legumes, it contains hundreds of diverse species that have nitrogen-fixing bacteria living symbiotically in nodules on the roots. Most species have compound leaves with rounded edges. Includes peas, chickpeas, many types of beans, clover, mimosas, and acacias.

FAGACEAE: Order of angiosperm dicots that are almost exclusively large, woody forest trees. Trees have separate catkin male flowers and simple female flowers on the same tree. Often produce nuts or samaras as seeds. Almost all depend on root mycorrhizae for healthy growth and most are temperate and deciduous. Includes maples, oaks, beeches, chestnuts, hickories, pecans.

FERTILIZATION: The fusion of male and female gametes of the same species, resulting in the development of a zygote that will undergo mitosis to create an embryo, and eventually, a new generate of adult sporophyte plants. Occurs in non-flowering plants when sperm cells swim to the archegonium of a female plant through water droplets. In seed plants, pollen sticks ot the pistil and grows a tube that allows sperm to swim in to the ova.

FIBROUS ROOTS: Type of roots most commonly employed by monocots and some dicots native to windy and/or areas with soil that easily erodes. Fibrous roots are

FILAMENT: Among angiosperm flowers, it is the stalk at the base of the stamen that lifts the anthers to the level of the pistil so that pollinators will spread pollen onto the female stigma as they feed on nectar.

FLOWER: In angiosperms, the reproductive organ that contains both male and female gametophytes inside the anthers and pistil, respectively (perfect flowers) or one type of gametophyte in flowers of separate sexes (dioecious imperfect flowers). The flower includes the petals and sepals, which are modified sporophyte leaves.

FOLLICLE: A type of dry dehiscent fruit that forms pods that split open only on one side at maturity, as opposed to legumes which open on both sides. Plants that form follicular fruits include peonies, delphiniums, and larkspurs.

FRAGMENTATION: A method of asexual reproduction, wherein pieces of a parent organism divide by mitosis to regenerate a new clone from undifferentiated tissues. Any plant that grows from a cutting uses this type of asexual reproduction, such as roses, cacti, coleus, and tomatoes.

FREE PLACENTATION: Found in unilocular fruits, this arrangement occurs when seeds attach to the lateral ovarian walls and float freely in the center of the fruit. Figs are a classic example of this type of placentation.

FRUIT: In angiosperms, a mature ovary containing fertilized ova (seeds). Numerous types of fruit have evolved. Some are dry and fibrous and protective of the seed embryos, while others others are fleshy

FROND: The leaves of the sporophyte generation of ferns, which are usually in a pinnate arrangement. Spore cases on the underside called sori divide meiotically to produce spores of both genders. Also the term for pinnate leaves in unrelated cycads (gymnosperms) and palms (monocot angiosperms).

FUNICULUS: A stalk that projects from the endoderm of the ovary (fruit) in plants that articulates at the hilum of the developing seed, anchoring it inside the fruit.

GAMETANGIA: Plant tissues that contain germ-line cells that divide meiotically to produce male or female gametes. In spore plants, these are ova and sperm, while gymnosperms and angiosperms enclose their sperm in grains of pollen.

GAMETES: Haploid reproductive cells produced meiotically by germ line cells in the megaspore mother cells of the ovary (ova) or the microspore mother cells of the anther (pollen). They are haploid in diploid plant species and they contain half the original number of chromosome copies in polyploid species.

GAMETOPHYTE: In organisms like plants that employ alternation of generations as their reproductive strategy, this is the haploid generation of plants that divide meiotically to produce male and female gametes (ova and sperm or pollen). In algae and primitive plants, they grow during times of drought (smaller than the sporophyte and use less resources) and produce gametes when water is abundant. Protected and cared for by sporophytes in higher plants.

GEMMAE: In liverworts, cup-like indentations of meristematic tissue on the surface of the leaves that divide by mitosis to generate small spheres of cloned tissue by budding. Rain or water droplets bounce these out of the parent plant, where they grow into a new clones of the parent.

GENTIANALES: An order of dicot plants that usually have oppositely-oriented leaves, symmetrical flowers, a single integument that connects many seeds to their fruits, and the biochemical ability to produce many different toxic alkaloids among the members of the taxa. Examples include jasmine, oleander, rosy periwinkle, and coffee.

GERANIALES: An order of dicot plants that includes mostly herbaceous plants, many of which are annuals. Their flower petals are not fused and they commonly produce dry capsule and schizocarp fruits. Examples of plants in this tax include geraniums, creosote, and coca bush.

GERMINATION: The emergence of neophyte plant embryo from the seed to grow into a seedling. The rupture of the seed coat and growth of the plant can be triggered by water, light levels, day length, soil temperature, fire, passage through the digestive tract of an animal, or many other environmental factors, depending on the species.

GIBERELLINS: A class of plant hormones that regulate a number of processes in plants, including germination of stems, vertical growth and lengthening, flowering, fruit development, and mobilization of sugars in the phloem.

GINKGOALES: A monotypic order of plants with only one living species, the Gingko biloba. All other members are extinct. Characterized by fin-shaped ribbed leaves called megaphylls, wood fiber composed of mostly tracheids, and separate male and female trees. Males have pollen cones, females produce a single uncovered seed in their cones after germination.

GLUCOSE: A monosaccharide with a hexagonal-shaped carbon ring structure and a formula of C6H12O6. It is isomers with many other simple sugars like fructose. Commonly produced by plants from two G6P molecules after the Calvin cycle. Usually, it is chained into polysaccharide polymers like starch, cellulose, or pectin.

GLYCOSIDES: A class of protective (and often toxic) secondary plant compounds biochemically derived from sugars with modified side chain groups. Examples include cardiac toxins produced by oleander and cyanide-generating amygdalin found in almonds, apple seeds, cherry pits, and peach pits.

GNETOPHYTA: A division of gymnosperm plants that diverged from conifers long ago in evolutionary history. Including plants such as ephedra, welwitschias, and gnetums, they are characterized by their production of secondary alkaloid compounds, obvious staminate cones in males, superficially fruit-like female cones, and wood fibers made of mostly tracheids. Commonly, they are desert plants, but sometimes tropical.

GRANA: A stack of chlorophyll containing thylakoid sacs inside the chloroplast. Functioning like solar panels, they trap light photons with pigments and use the energy to run electron transport chains that make ATP and NADPH in the light reactions of photosynthesis.

GRAVITROPISM: Tendency of a plant to grow directionally under the influence of gravity. Occurs because signal molecules trigger receptors in plant cells that send secondary messages inside the cells, resulting in production of auxins and other hormones that cause cell division and growth of stems upward and roots downward.

GROUND TISSUE: A multi-purpose type of tissue in plants that starts out as non-descript undifferentiated cells that can develop into a diverse number of specializations. They can be fated to become parenchyma (softer, live, few cell walls), collenchyma (fibrous, live, moderate number of walls), or sclerenchyma (hard, dead, many cell walls).

GROWTH RINGS: In woody dicots, these are concentric layers of vascular tissue laid down each season by the vascular cambium. Spring wood tends to form lighter rings because cell growth is faster and the wood is less compressed, while Summer wood tends to form the dark rings, as growth slows down in hotter, drier conditions.

GUARD CELL: Pairs of cells that regulate the opening and closing of stomatal openings of leaves and green stems. By pumping potassium ions into the cells, membrane pumps can cause water to follow osmotically, swelling the cells, which are tethered together, which opens the stomata so that the plant can gather CO2.

GUNNERALES: A relatively primitive order of dicots that includes the gunnera plants native to the Tropical Americas. In addition to gigantic fan-shaped leaves and huge succulent petioles, they have simple two-part flowers, and unique symbiotic relationships with cyanobacteria that live on the scales of their stems and fix nitrogen.

GYMNOSPERM: A clade of plants that were the first to develop naked seeds inside of cones. They lack fruit or flowers. All are woody, perrenial plants and most are evergreen. They were the first group of plants to develop pollen and seeds, which allowed them to colonize much further onto land. Wood fiber is made of mostly tracheids. Leaves are usually needles or scales. Includes cycads, gnetophytes, ginkgos, and conifers.

HAPLOID: Refers to cells that contain only one copy of each chromosome in the genome. In spore plants, the gametophyte stage and associated ova and sperm are haploid. In higher plants, only the germ line gametophyte cells (microspore or megaspore mother cells) and the ova and pollen sperm are haploid.

HEARTWOOD: The dense, non-living primary xylem tissue found at the center of a tree trunk. It has been crushed and compressed by the outer layers and gravity over time. It is often rot-resistant, as it is impregnated with tannins and very hard lignin fibers. Important in support of large trees.

HEAT SHOCK PROTEINS: A class of proteins that is responsible for protecting other sensitive proteins from denaturation during times of temperature stress. Also called chaperone proteins, they cluster around enzymes and other critical proteins and interact with them in a way that prevents the sensitive protein from unfolding.

HELIOTROPISM: Tendency of a plant to grow toward sunlight and follow its path across the horizon. Occurs because phytochromes and other such molecules are triggered, sending secondary messages inside the cells that result in potassium ions being pumped in and out of stem cells, allowing the plant to change turgor pressure on the correct side of the stem to bend toward the sun.

HERBACEOUS: Any angiosperm plant that lacks secondary growth in the stem and roots. Rather than having woody trunks, they retain green stems with a dermal layer and a layer of ground tissue (cortex and/or pith) inside the roots and stems. The lack a cork cambium or bark. Examples include tomato plants, daisies, lilies.

HERMAPHRODITE: An organism that has both male and female reproductive systems. Ferns, conifers, and the majority of angiosperms possess both male and female organs and are capable of producing sperm from the antheridium or inside the pollen in the stamens and ova inside the archegonium or pistil.

HESPERIDIUM: A fruit that is actually a large berry with a mesocarp divided into segments, surrounded by a tough leathery rind or ectocarp. This type of fruit is produced by all citrus species and a few other obscure tropical plants.

HETEROSPOROUS: Refers to spore plant species (such as moss) and algae species, wherein two different genders of spores develop meiotically from different groups of germ-line cells. Small microspores produce male plants, while larger megaspores produce female plants.

HILUM: The circular scar on a seed where the funiculus attached the seed to the ovary.

HOMOSPOROUS: Refers to spore plant species (such as horsetails) and algae species, wherein two different genders of spores develop from spores that are physically almost indistinguishable in size and appearance.

HYPANTHIUM: Refers to flowers that have a cup-shaped basin between the stem and the ring of petals above it.

HYPOCOTYL: In plant embryos, this stalk-like part of the neophyte is located between the epicotyl and cotyledons (embryonic leaves) and the radicle (root) fated to become mostly the stem.

HYPOGYNOUS PISTIL: Refers to flowers that have pistils elevated well above the level of the pedicel, so that the pistil sticks out of the middle of the flower conspicuously, with the base of the ovary exposed.

IMPERFECT FLOWER: Refers to an angiosperm flower that has one specific set of gender organs for reproduction. Species that make them produce two types of flowers on the same plant or may grow as totally separate male and female plants. Examples include pomegranate, squash, and kiwi fruit flowers.

INDEHISCENT DRY FRUIT: Refers to types of dry fruit that do not have a natural seam for seeds to escape. For the seeds to get out, the seed coat either must be split open some way. This can be by swelling in water, fire, or by having animals gnaw it open. Examples include Brazil nut pods, acorns, maple seeds, and sunflower seeds.

INFLORESCENCE: Refers to the acte of flowering, or to the specific type of arrangement of flowers on a plant.

INTEGUMENT: The fleshy or fibrous part of the endocarp that directly forms an attachment from the ovary (fruit) to ovules (seeds) via the funiculus and hilum. For instance in a jalapeno pepper, this is the fibrous white rib in the middle of the fruit that holds all the seeds.

INULIN: A storage polysaccharide that is an alternative to starch in certain types of plants. It is a polymer of fructose, rather than glucose. In monocots, many members of Poales store inulin almost exclusively, such as garlic, while in dicots, members of Order Asterales, such as asters and daisies, also do so.

K-STRATEGIST: Refers to an organism that puts most of its reproductive strategy into maintaining a population at the carrying capacity as opposed to trying to max out reproductive rate. Typically these organisms have long lifespans, high investment of resources, and high survival rates. This strategy is not applicable to most plants, but large forest trees use this life strategy if they survive the seedling stage.

LATERAL BUD: Meristematic tissue located on the stem that generates horizontal branches and leaf growth. Also known as axillary buds, they increase the bushiness of a plant at the expense of terminal lengthening.

LAMIID FLOWER: Refers to angiosperms that develop their flower petals from distinct lumps of tissue in the embryo, rather than a concentric layer of tissue, as it does in campanulid flowers.

LAMINALES: An order of dicot angiosperms that have perfect lamiid flowers, opposite leaf arrangements, and two-chambered ovaries. Many members of this order are herbaceous or shrubby and produce aromatic secondary compounds. Includes such plants as mint, basil, lavender, olives, and lilacs.

LATERAL MERISTEM: A region of undifferentiated cells that results in horizontal and axillary growth in a plant. They produce axillary branches and buds and lateral roots, but deeper layers of meristem in the trunks and roots of woody plants divide to produce new layers of xylem, phloem, and cork and increase the tree's girth.

LEAF: The primary photosynthetic organ of most plants. Consist of a sandwich of photosynthetic parenchyma tissues and vasculature between wax-covered layers of epidermal tissues. Evolved from tissues that emerged from photosynthetic stems in primitive plants, such as club mosses.

LEAF SCAR: An impression or indentation left on the stem of deciduous dicot trees by the prior season's leaf petiole before senescence caused the leaf to be shed.

LEAFLET: In compound pinnate leaves, the smaller modular leaves that emerge laterally from the main petiole or from a secondary petiole. Examples include the leaves of hickories, sumacs, and sweet peas.

LEGUME: A member of Order Fabiales that is capable of fixing nitrogen for the synthesis of proteins and nucleic acids. It acquires nitrates symbiotically from various bacterial species that it hosts inside nutrient-rich root nodules.

LEUCOPLAST: A type of plastid found in many plant cells that is derived from proplastid tissues and is found in non-photosynthetic tissues, such as the cortex of the stem and roots. Their primary function is the storage of starches, complex carbohydrates, and lipids.

LENTICEL: A small opening found on the surface of woody stems and trunks that provides a means of exchange for carbon dioxide, oxygen, ethylene, and other gases.

LIGHT REACTIONS: The set of reactions conducted by the electron transport chains of photosystems I and II in the thylakoid membranes of chloroplasts to generate ATP and NADPH for use in the Calvin cycle. Water is split and its protons and electrons are separated by the membrane, creating an electrochemical gradient that allows energy to be generated. Light is necessary to excite the electrons for later energy transfer.

LIGNIN: An extremely complex organic polymer made of heavy alcohols and aromatic compounds that is produced in the xylem cell walls of woody plants, especially in the trunk and roots. In addition to its great strength as a weight support tissue, it cross-links to cellulose in cell walls of vascular tissue, aiding in material transport.

LILIALES: An order of monocot angiosperms that have prototypical three-part flowers that don't produce nectar, with the petals and sepals often being hard to distinguish, flat blade-like leaves with parallel leaf veins, and tuber or corm-like roots that may reproduce asexual by budding. Includes all types of lilies, tulips, crocus, and smilax brier.

LIPIDS: A class or biochemical compounds that are non-polar or mostly non-polar, rich in carbon-hydrogen bonds, and used for a variety of purposes. Classes include triglycerides (fats and oils), waxes (waterproofing), sterols (hormones and cell membranes), phospholipids (cell membranes), pigments (energy absorption), and others.

LOCULE: Refers to a divided compartment within the ovary of a plant. Some plants have monolocular ovaries, while others can have as many as six subdivisions that each ovulate separate ovules.

LONG DAY PLANT: A type of plant that is triggered to flower by phytochromes exposed to long hours of sunlight near the Summer equinox. Plant will not bloom with too many hours of darkness. Examples of these include potatoes, asters, poppies, lettuce, and many other plants native to cooler temperate areas that need to be certain they will have time to develop seeds before frost.

MACROALGAE: Any type of algae that superficially resembles a plant, such as kelp (brown algae), sea lettuce (green algae), or Irish moss (red algae). While their leaf-like thalli and stem-like stipes resemble plant parts, they are not, due to the absence of a vascular system, differing pigment mix, and biochemically different storage compounds.

MACRONUTRIENT: A nutrient required by plants in mass quantities, such as C, H,O, N, P, S, K, Ca, Mg, and a few others. All of these elements are essential for the construction of organic compounds and/or the proper function of certain enzymes, pigments, or ongoinog homeostatic processes in cells.

MALIC ACID: An organic acid formed by the Kreb's cycle and after the uptake of CO_2 at night by C4 plants in the bundle sheath cells. Later, CO_2 is cleaved off of the 4 carbon acid, allowing the Calvin cycle to run in the chloroplasts without the need to open the stomata during the day and risk water loss.

MALPHIGIALES: A large and diverse order of dicots that is united by cellular biochemical similarities, DNA sequence evidence, and flower morphology. Their flowers lack nectaries and, instead, offer bees wax. Many produce toxic secondary compounds. Members include willows, passionfruit, euphorbias, castor beans, and poinsettias.

MALVALES: A large order of dicots commonly called mallows. They have toothed or palmately-lobed leaves with deep, defined veins, trumpet-shaped flowers with a fused calyx, and fibrous tissues or woody tissue in the stems. They also produce mucilaginous compounds and have leaf hairs to deter herbivores. Members include okra, hibiscus, durian fruit, African baobob trees, cotton, and cacao.

MALVID CLADE: A very large group of dicots that represents about 25% of known species. DNA phylogeny has been used to work out the relationships and likely evolutionary patterns. Includes many commonly known orders and families, including the mallows, the cucurbits, geraniums, myrtles, citrus, and brassicas.

MARGINAL PLACENTATION: Found in types of fruit where the seeds attach toward the lateral edges of the ovary. Found in beans, peas, brassicas, and lupines.

MEGAPHYLL: Like those found in all dicots, large sheet-like leaves that bridge gaps between a complex network of veins with large surface areas. Also, a really big guy whose full name is Phyllip.

MEGASPORE MOTHER CELL: A diploid germ line cell in the ovary of plants that divides by meiosis to give rise to haploid megaspore cells that then divide by mitosis to make up egg cells, synergids, and other embryo parts.

MEGASPOROPHYLL: Leaf or stem-like appendages that protectively house megaspore mother cells in gymnosperms. Comparable to a pistil in flowering plants. In conifers, megasporophylls make up the spiky petals of the female cone that protect the seeds after the embryo develops.

MEMBRANE TRANSPORT PROTEINS: Proteins embedded in the cell membrane that regulate the passage of materials according to intermolecular attractions and shape and size matches (carrier proteins) or by charge and diameter (ion-channel proteins). Other than gases, most substances cannot penetrate the phospholipid bilayer and most go through the proper 'door' to gain entry or exit.

MEIOSIS: Process of cell division that generates four haploid gametes (eggs, sperm, spores) from diploid germ-line cells by dividing twice through meiosis I and II. Meiosis I allows homologous chromosomes to pair up, crossover, trade genetic material, and be shuffled at random into two new cells, while meiosis II creates haploid cells by splitting sister chromatids in half and dividing the cell again.

MERISTEM: An area of growing undifferentiated tissue in plants that is not yet fated to become a specific type of cell. Similar to stem cells in animals, they can divide to become whatever tissue type might be needed, eventually switching genes on and off until they have changed into a specific tissue type.

MESOCARP: The middle layer of tissue in a fruit that often forms the fleshy, edible layer, such as in an apple or a peach. Some mesocarps, however, can be dry and fibrous or be impregnated with toxins to protect the seeds.

MESOPHYLL: Layers of ground tissue in plant leaves that are usually composed of cells packed with chloroplasts for the purpose of food production via photosynthesis. They typically have relatively few cell walls to allow materials to be exchanged quickly. Monocots have a single dispersed layer, while dicots have a layer of packed palisade mesophyll and a layer of spongy mesophyll with air spaces between the cells.

MICRONUTRIENT: A nutrient required by plants in trace amounts, usually as a cofactor in enzymes, such as those in photosynthetic and metabolic pathways. Excess amounts of many of these can be toxic to the plant. Examples are elements such as Cu, Co, B, Fe, Mn, and Ni.

MICROPHYLL: Small, simple leaves that have a single leaf vein. Essentially, they are photosynthetic tissue that evolved to expand the photosynthetic surface area of green stems. Examples are the simple spike-like leaves seen in club mosses and horsetails.

MICROPYLE: The opening on the surface of an ovule that allows a pollen tube to penetrate so that sperm can swim through the tube for fertilization. In developed seeds, it can often appears as a pinhole-like opening.

MICROSPORE MOTHER CELL: A diploid germ line cell in the anthers of plants that divides by meiosis to give rise to haploid microspore cells that then divide by mitosis to make up sperm cells, a tube cell, and a pollen coat.

MICROSPOROPHYLL: Leaf or stem-like appendages that protectively house microspore mother cells in gymnosperms. Comparable to a stamen in flowering plants. In conifers, microsporophylls make up the leafy petals of the soft male cones that release pollen grains.

MIDRIB: Originating at the petiole, it is the main vessel extending into the blade of a leaf, carrying water into the leaf from the xylem in the stem or trunk and transporting manufactured sugars into the phloem to be carried away.

MITOSIS: A type of cell division that results in two daughter cells that are exact clones of the parent cell with the same ploidy. Used for growth and repair in plants, but also in asexual reproduction in plants that can perform budding or fragmentation to create clones.

MONOCARPELLARY FLOWER: A flower that has a single simple pistil in the middle of the corolla. Examples are the flowers of mangoes and legumes.

MONOCOT/MONOCOTYLEDON: A clade of angiosperms that appears to have evolved from basal dicots in the distant past. They are characterized by seeds with a single cotyledon, flower parts in multiples of 3, scattered vascular bundles in the stem, parallel leaf veins, fibrous roots, and other commonalities. Examples include palms, ginger, lilies, banana trees, agaves, asparagus, grasses, and bromeliads.

MONOSULCATE POLLEN: Pollen grains that have a single ridge in the middle of their protective coat. This trait is a common characteristic of basal dicots, such as avocados, and more primitive types of monocots, such as irises. This lends credence to the idea that monocots evolved from this branch of dicots.

MUTUALISM: A symbiotic ecological relationship where two species live in close association, assisting each other by trading one or more useful benefits or another. Often these pairs of species have co-evolved together. In plants, pollinators and angiosperms have this relationship, as do mycorrhizal root fungi and trees.

MYCORRHIZAE: Mutualistic fungal species of phyla Basidiomycota (club fungi) and Ascomycota (sac fungi) that grow mycelial networks over and into the roots of plants, exchanging additional surface area for water and soil mineral uptake, in exchange for sugars and other nutrients from the plant.

MYRTALES: An order of angiosperm dicots that are mostly (but not always) woody shrubs and trees. They have a double layer of phloem around the wood xylem and often produce aromatic compounds in their sap as an insect deterrent. Some members of this order exhibit polyploidy. Fruits are usually berries. Includes myrtles, eucalyptus, pomegranates, and primroses.

NADPH: An electron-carrying coenzyme produced in the electron transport chain of photosystem II in the thylakoid membranes of chloroplasts. After the formation of ATP, it forms from NADP+ and the electrons and proteins collected at the end of ATP formation. NADPH is ferried on to the Calvin cycle in the stroma. Enzymes there attach the hydrogen atoms onto the carbon skeleton of sugars and NADP+ returns to the thylakoids.

NECTAR: A concoction of various sugars and aromatic compounds offered as a mutualistic food source to pollinators in exchange for the transfer of pollen from the stamens of the flower to the pistils of other individuals. The location of nectaries varies widely between the flowers of different taxa.

NEEDLE: A type of microphyll leaf found in gymnosperms such as cycads and conifers.

NERIOSIDES: Toxins manufactured by certain members of the Order Gentianales, such as oleanders and foxgloves, which interfere with acetylcholine receptors, ATPase enzymes, and ion channels of animals that might try to consume them. This causes potentially lethal heart arrythmias, kidney problems, and other toxic effects.

NITRIFICATION: The metabolic process performed by certain soil bacteria that results in the production nitrates after the bacteria splits the protons and electrons off of ammonia to run its electron transport chains during ATP production. Plants benefit from this process, as nitrates are soluble and can be taken up as a nitrogen source.

NITROGEN CYCLE: The biogeochemical cycle that results in the conversion of nitrogen into various different compounds in steps mostly governed by bacteria. Atmospheric nitrogen gas passes through nitrogen fixation, and nitrification before being taken up by plants and animals. Wastes and dead organic matter are returned to the soil and converted by ammonification bacteria, with denitification bacteria regenerating atmospheric nitrogen.

NITROGEN FIXATION: A step of the nitrogen cycle that results in the conversion of inert atmospheric nitrogen into soluble ammonia by bacteria that use it as a final electron acceptor. Fixation locks the nitrogen in the soil for later steps that make nitrogen available to plants.

NUCLEIC ACIDS: A class of biomolecules composed of nitrogen-rich ring-shaped molecules called purines and pyrimidines that are linked to a monosaccharide and phosphate backbone. DNA, RNA, ATP, and NADPH all contain nucleic acids in their structure. Plants need a source of soil nitrogen to manufacture them.

NUT: An indehiscent dry fruit that contains a hard sclereid-rich seed coat protecting a nutrient-dense single seed. They are the dry fruit equivalent of a drupe. True nuts include acorns, hickory nuts, and mint seeds.

OOCYTE / OVA: A (usually) haploiId female gamete produced meiotically from the megaspore mother cell germ line inside of the ovary of plants. It will join with sperm from the pollen to create an embryo inside of a seed.

OSMOSIS: Diffusion of water across a selectively permeable membrane through tiny donut-shaped membrane proteins called aquaporins. Dissolved substances are blocked by the membrane. As a result, water will move across a membrane toward the area of higher solute concentration until equilibrium is reached. Osmosis regulates turgor pressure inside the central vacuole of plant cells, keeping herbaceous plants and leaves swollen and supported.

OVARY: The female reproductive organ that contains the megaspore mother cells that divide meiotically to produce ova. Located inside the archegonium of seeRdless plants, the cones of gymnosperms, and the flower pistils of angiosperms. In angiosperms, it matures into a fruit.

OVULE: Gametophyte follicles inside of a plant ovary that are fated to become seeds containing the embryo, endosperm, and seed coat. A mature fruit is the ovary and the mature ovules are seeds.

PALISADE PARENCHYMA: In C3 plant leaves, a tightly-packed layer of rectangular ground tissue cells that are loaded with chloroplasts. Their primary function is production of biomolecules via photosynthesis.

PALMATE LEAF: A compound leaf that has fused leaflets in a star, hand, or fan shape emerging from the main petiole. Examples include Virginia creeper, sweet gum, marijuana, schefflera and various species of fan palm.

PANDANALES: An order of dicot angiosperms that are unusual among monocots in having four stamens and two pistils in the flowers. They have very large endosperms in their seeds. Leaves have parallel veins and roots are fibrous. Includes diverse array of plants including pandanus, Panama hat plant, duckweed, and arums.

PARAPHYLETIC: Refers to an evolutionary classification that groups organisms together erroneously, since molecular evidence has revealed them to not be closely related, but superficially similar. Usually this occurs because classification systems the predated DNA technology have used similar morphologies to classify organisms. Kingdom protista is a classic example, as most algae taxa are only superficially similar, but totally unrelated.

PECTIN: A polysaccharide produced by plants that act as a binding agent that glues secondary cell walls onto prior cell walls and to neighboring cells. It is composed of repeating units of galactouronic acid.

PEDICEL: The connecting stalk of tissue that links a flower to the stem of a plant.

PENTALOCULAR OVARY: An ovary that uses the ovarian wall to form five distinct divided lobes with separate seed compartments. A common type of arrangement in mallows like okra and cotton in Order Malvaceae.

PENTAPETALIDS: Angiosperm dicots that have evolved distinctly different flower petals and sepals in multiples of five. This is generally considered to be a more advanced derived evolutionary trait versus dicots with four petals.

PEP CARBOXYLASE: A very fast and efficient enzyme found in C4 plants that traps CO_2 from the atmosphere when the stomata are open (usually at night) and passes it on to a biochemical pathway that creates stored malic acid in the bundle sheath cells. During daytime, the stomata remain closed, while CO_2 is cleaved off of malic acid for use in the Calvin cycle in the parenchyma.

PEPO: Essentially a giant berry with multiple seeds and a tough rind. Examples include cucumbers and pumpkins.

PERFECT FLOWER: A hermaphroditic flower that has a complete set of both male and female reproductive organs, giving it the capability to produce both viable pollen and ovules.

PERICARP: The part of a ripened fruit that originates as the wall of the ovary at the base of the flower pistil. For example, the peel, flesh, and core of an apple are the pericarp that surrounds the seeds.

PERICYCLE: Meristematic tissues located on the outer perimeter of xylem, phloem, and procambium cells in roots. This tissue divides and differentiated to produce lateral roots.

PERIDERM: In woody plants, the outer layer of cork produced by the cork cambium, which forms bark that splits and replaces juvenile dermal tissues in mature trees.

PERIGYNOUS PISTIL: Refers to flowers that have an elevated stigma above the level of the pedicel, but whose ovaries remain tucked away inside the pedicel for protection.

PERIPHERAL PLACENTATION: A seed arrangement found in fruits that have a central dividing pith created from modified ovary wall tissues, but the seeds attach laterally to the mesocarp instead of at the central axis. Found in pumpkins, squash, cucumbers, and melons.

PERRENIAL: A type of angiosperm plant with an indeterminate lifespan. As opposed to short-lived annuals and biennials, perrenial plants bloom and produce seeds on a regular cyclical basis over the course of their lifespans.

PETAL: A modified sporophyte leaf that surrounds the peduncle at the base of a flower. Some petals are merely protective sheaths, such as in grass, while others are modified in shape and pigmentation to attract pollinators.

PETIOLE: The vascular stalk that connects a leaf and a stem. It is a conduit that carries water from the xylem into the tissues of the leaf and transports manufactured sugars and other photosynthetic products to the stems and roots.

PHLOEM: Living vascular tissues found throughout plants that are responsible for the transduction of sugars. They are derived from meristematic tissues of the vascular cambium. Comprised of hollow donut-shaped seive tube elements that use membrane transport proteins to pump sugars, along with companion cells that care for them.

PHOTOPERIOD: The prescribed period of daylight that acts as a stimulus for various homeostatic processes in plants. In many plants, the ratio of daylight to dark (and the intensity of the light) triggers hormonal changes that influence many processes related to senescence, flowering, and growth.

PHOTOSYSTEM I: The second stage of the light reactions of photosynthesis. Paradoxical name because it was the first step of the light reactions discovered. Its reaction center contains mostly chlorophyll B, which receives energized electrons from photosystem II via the electron carrier plastoquinone. Its electron transport chain generates ATP and NADPH for use in the Calvin cycle.

PHOTOSYSTEM II: The first stage of the light reactions of photosynthesis. The hydrolase enzyme at its inception splits water into protons, electrons, and oxygen. The electrons are picked up by mostly chlorophyll A, which traps photons of light that energizes them. The electrons jump off the pigment molecule down a series of electron carriers that carry them on to photosystem I.

PHOTOTROPISM: Tendency of a plant to grow directionally toward incoming light. Phytochrome molecules respond to the light and trigger secondary messenger pathways that ultimately result in production of auxins and other hormones that cause disproportionate cell division and growth of stems on one side of the plant.

PHYTOCHROME: Pigments found in the cytoplasm of plant cells that are composed of a porphyrin ring component combined with a protein component. There are five known types of these pigments that are all most sensitive to slightly different wavelengths of light. Light causes configurational changes in proteins that interact with DNA transcription factors, turning various genes on or off.

PINNATE LEAF: A compound leaf arrangement that has perpendicular leaflets flanking a large central petiole. Examples include Christmas ferns, coconut palms, and hickories. Some leaves are twice-pinnately compound, with smaller petioles with leaflets emerging from a main petiole, such as in mimosas and locust trees.

PISTIL: The main female reproductive organ at the center of angiosperm flowers. Includes an ovary at its base, which narrows into a bottle-shaped style tipped by a sticky pad known as a stigma. The stigma traps pollen, the style allows the growth of a pollen tube, and the ovary matures into the fruit after the ovules are fertilized and develop into seeds.

PITH: A starchy and fibrous layer of cortex tissue at the center of an herbaceous dicot plant. Used for food storage and structural support. Examples of plants grown for the pith include rhubarb and kohlrabi.

PLACENTATION: Pertains to the position of the hilum and funiculus that attaches the seed to the fruit. There are multiple types of placentation, depending on whether the seeds attach in the middle to the outer wall of the ovary, or in several other orientations. One of many reproductive clues that offer a means of plant classification.

PLASMODESMATA: Openings in plant cell walls that form cytoplasmic tunnels between the extracellular spaces of neighboring plant cells for membrane transport. Cell walls block much of the membrane transport that is possible in the extracellular spaces, as in animal cells.

PLASMOLYSIS: Refers to the shriveling of a plant cell when exposed to a hypertonic solution. High solute concentrations on the exterior cause water to travel out of the cell via osmosis, resulting in the emptying of the central vacuole and collapse of the cell wall onto the cell. Collectively, this will cause the entire plant to wilt.

PLASTID: Any of a number of membrane-bound plant organelles derived from protoplastids. They function in photosynthesis (chloroplasts, chromoplasts, etc.) or food storage (leucoplasts, amyloplasts, etc.).

POACEAE: A very large order of angiosperm monocots colloquially called 'grasses', though many other plants also belong to this order. Characteristics include unremarkable bract-like petals (except bromeliads) around clusters of small flowers, seeds with a large starchy endosperm, jointed stems, parallel leaf veins, fibrous roots, and (often) C4 metabolism. Includes grasses, maize, wheat, barley, oats, rice, bromeliads, sugarcane, and bamboo.

POLAR NUCLEI: Two gametophyte nuclei that fuse to become a diploid cell after two rounds of mitosis in a developing gametophyte. It is fated to be fertilized by one of the two sperm that enter the ovary through the pollen tube. It ultimately becomes the nutritive endosperm that stores starch and nutrients inside the seed.

POLLEN: The male gametophyte in plants, which forms after meiotic division inside the tissues of the male cone or anther of a flower.. A pollen grain contains a protective coat, two sperm cells, and a tube cell that provides a conduit into the ovary and eliminates the need for an external water source for the sperm to swim to the egg.

POLLINATOR: A mutualistic animal partner of a plant that provides reproductive services by transferring pollen grains from the anthers of one plant to the pistil of another. The plant offers an exchange of nectar and/or protein-rich pollen. Common pollinators include honeybees, butterflies, flies, wasps, bats, hummingbirds, and marsupials.

POLLINATION: The transfer of pollen from the male staminate cone or anthers of one plant to the female pistil of another plant. During the process of pollination, the pollen coat glues itself to the sticky secretions of the stigma and opens to release the tube cell. The tube cell grows down the style into the base of the pistil until it contacts an ovule, at which time the sperm swim down the tube for fertilization.

POLYCARPELLARY FLOWER: A type of flower with four or more ovaries within the corolla. Allows for the production of numerous seeds from a single flower. Examples are carnations and poppies.

POLYPLOID: Refers to a cell or organism that has more than two sets of identical homologous chromosomes in the nucleus/nuclei. Polyploidy is common in plants, since more copies of genes can compensate for the reproductive limitations caused by immobility. Tetraploidy (4n) (bananas), hexaploidy (6n) (oats), and octoploidy (8n) (strawberries) are all relatively common.

PORPHYRIN: A class of biomolecules that are manufactured from glycine and succinic acid, which has an intricate nitrogen-rich structure composed of interlocking carbon rings. Porphyrins are used for oxygen transport, but are also used in the structure of many pigments, since the many electrons in their double bonds can be excited. Chlorophylls belong to this class.

PRIMARY GROWTH: Refers to vertical growth in plants that emerges from the apical meristems at the shoot tip and the root tip.

PRIMARY MERISTEM: Undifferentiated tissues at the shoot tip and root tip of a plant that can divide to produce new tissues of any type needed to form new roots, stems, and leaves.

PRIMARY PHLOEM: In primary growth, it refers to the newest phloem cells that emerge from apical meristem tissues. In secondary growth, it is the newest layer of phloem that emerges on the exterior side of the vascular cambium in the stem and roots. Also refers

PRIMARY PRODUCTIVITY: A measure of the food production capacity of a given habitat, which dictates the potential size of the food web. It directly relates to the amount of photosynthesis per given area per year. Typically, it is measured in kCal/m2/yr or as g Carbon fixed/m2/yr.

PRIMARY XYLEM: The first layer of xylem that emerges from the vascular cambium toward the center of a stem or root. In spite of being the newest layer of xylem that is the most effecient at transporting water, it is dead at functional maturity, with numerous layers of cell walls creating 'water pipes' that are hollow in the middle.

PROCAMBIUM: The layer of meristematic tissue that divides and differentiates into vascular tissues. The vascular cambium of the stems and roots is derived from procambium, but procambium tissue is also found in leaves and at shoot and root tips.

PROMOTER: A DNA sequence upstream of a gene that acts like a control box mechanism. Its sequence is recognized by DNA polymerase and by transcription factor proteins that control the rate of messenger RNA synthesis. In plants, secondary messengers triggered by hormones and phytochromes influence the transcription factor proteins that interact with genes.

PROTONEMA: In bryophyte plants, this is the early juvenile stage of the gametophyte. It divides repeatedly by mitosis to mature into the adult male or female plant.

PTEROPHYTA: A phylum of prehistoric and primitive plants that are only marginally more evolved than green algae. Includes mosses, liverworts, and hornworts. They lack a vascular system, have primitive rhizoids and no true roots, have primitive microphyll leaves, and rely on water to carry sperm to the ova. They produce spores instead of seeds and have no cones, flowers, or fruit.

PYRUVIC ACID: The 3-carbon end product of glycolysis in cellular respiration that is passed on to the Kreb's cycle for the next step of ATP generation in cellular respiration. Is also generated by C4 plants when they split the 4-carbon malic acid produced by PEP carboxylase to release CO2 for the C3 pathway of photosynthesis.

R-STRATEGIST: An evolutionary reproductive strategy, wherein the species in question tries to maximize the reproduction rate at the expense of survival odds. They typically produce a plethora of offspring, invest little or no parental care, and have short lifespans. The majority of plants, especially herbaceous, use this strategy.

RACEMOSE FLOWERS: An inflorescence arrangement where numerous lateral stalks of flowers grow from a main spire. Typically, racemose flowers are small and numerous. Examples include mustard, crepe myrtles, and mint.

RADICLE: The first embryonic root that emerges in development from the apical meristem. It breaks the seed coat during germination, providing the first anchor into the soil and the first means of nutrient uptake.

RECEPTOR PROTEINS: Found on the surface of cells and embedded in the membrane, these proteins receive molecular signals from other cells or external stimuli. They trigger changes inside the cell via a system of secondary messnengers that ultimately turn genes on to produce proteins required for the prescribed changes.

RHIZOID: In bryophytes, small root-like projections that anchor the plant into the soil or onto a substrate. They are not true roots because they lack a vascular system.

RHIZOME: An underground stem that grows laterally, helping the plant spread by budding and drastically increasing the surface area for water and nutrient uptake. Meristematic tissues at the surfaces generate small vertical roots that penetrate the soil and regenerate new stems and leaves at the top. Rhizomes are the main anchoring system used by pterophytes, but are also common among angiosperms.

RIBULOSE BIS-PHOSPHATE: In the C3 pathway, this is the 5-carbon molecule that cycles back from the prior Calvin cycle to receive another molecule of CO_2 when the Rubisco enzyme traps it from the air. The resulting 6-carbon molecule immediately splits and the two 3-carbon products (3-phosphoglycerate) re-enter the Calvin cycle.

ROOT: The primary organ used for water uptake, nutrient absorption, and food storage in most plants. Composed of 3 different types of tissues. A layer of dermal tissue on the outside extends into asborptive root hairs and may host nitrogen-fixing bacteria colonies. A layer of ground tissue may be present to store water and food. An inner layer of vascular tissue contains the xylem and phloem that conduct water and sugars.

ROOT CAP: The apical meristem on the very tip of a root that penetrates the soil and divides to generate longer roots. It is often tipped with a tough dermal layer for protection.

ROOT HAIRS: Microscopic extensions of the root epidermis that increase the surface area for water and nutrients absorption by thousands of times. Frequently also used to host beneficial mycorrhizal fungi and nitrogen-fixing bacteria colonies.

ROOT NODULES: Tumor-like swellings on the roots of lemuge plants in Order Fabales, alders, and a few other tropical plants that host nitrogen-fixing bacteria. In exchange for the nitrogenous compounds they generate, the bacteria colonies are offered a sheltered home with access to biochemical nutrients.

ROSALES: Order of angiosperm dicots that tend to be woody shrubs, vines, and small trees. They have stipules at the base of their leaves, whorled cup-like flowers with a single pistil and numerous stamens, and usually produce drupe or pome fruit with hard stone-like seeds full of cyanogenic toxins. It is a large order than includes roses, apples, peaches, cherries, many other fruit trees, marijuana, and elms.

RUBISCO: The first enzyme of the Calvin cycle that fixes carbon by combining incoming CO2 molecules from the stomata with the 5-carbon RuBP sugar from the prior Calvin cycle to produce a 6-carbon sugar. The 6-carbon sugar immediately splits into two molecules of 3-phosphoglycerate, trapping the carbon resources in the plant.

SALINIZATION: The process that occurs in the soil when excess nutrients lower the water potential of the soil to the point that it draws water out of roots osmotically. A frequent problem caused by overfertilization or an environmental pressure found in estuaries and deserts that salt-tolerant plants have adapted for.

SAMARA: A wing-like dry indehiscent fruit similar to an achene, with a functionality of distributing its seed away from the parent on the wind, rather than providing a nutritional bribe to an animal. This strategy is used by maples, elms, and ashes.

SAP: The solution of dissolved sugars and nutrients that circulates through the phloem tissues of a plant. Membrane transport pumps mobile sap from the roots during Spring in many trees to assist in leaf development.

SAPINDALES: An order of angiosperm dicots that are mostly woody shrubs and trees. Alternately arranged compound leaves are the norm, though there are exceptions. Members of this group have glands in their leaves that produce aromatic oils to deter herbivores. Thorns are common among the group. Many members of this group are at least partially dioecious. Includes citrus, sumacs, poison ivy, cashews, lychee, and mangoes.

SANTALALES: An order of angiosperm dicots considered to be primitive, as compared to other members of the asteralid clade. They have no seed coats and less chlorophyll in their leaves, with some evolving to be parasites of other plants. Fruits are usually berries. Members of the order include sandalwoods, mistletoe, and witches broom.

SARCOTESTA: A foul-smelling, fleshy cone-like structure that surrounds the seeds of female gingkos.

SAXAFRAGALES: An order of woody angiosperm dicots whose members show molecular similarities in their DNA and RNA sequences, but whose morphologies vary greatly. Dry follicular fruits are often seen among the group,, as are star-shaped leaves. Flowers are often racemose and typically have two pistils surrounded by stamens. Membes include sweet gums, stonecrops, alders, gooseberries, and astilbe.

SCALE: Can refer to the flat, fibrous, turpene-impregnated microphyllic leaves of many conifers, such as spruces and junipers or to the protective flat, dry clusters of fibrous leaves surrounding bulbs and rhizomes in angiosperms.

SCHIZOCARP: An indehiscent dry fruit that grows in clusters, but splits apart when the seeds are distributed. Examples include dandelions, parsley, carrots, and dill.

SCLEREIDS: A hard, fibrous type of sclerenchyma ground tissue that contains many layers of cell walls. Functionally dead at maturity, they form bundles of fibers and stony growth of several different shapes. Examples include the seed coats of peach pits, beans, and coconuts, the fibrous part of bark in many different trees (such as cinnamon sticks), and as support for stems in monocots with false trunks, such as palms.

SCLERENCHYMA: A type of ground tissue that is primarily used for strengthening surrounding tissues against damage or gravity. Multiple lignified cell walls grow around these cells until they are hard and rigid. They provide support rods for vascular tissue, coverings for seed coats, and gravitational support in stems.

SCLEROTESTA: The hard, lignified coat that surrounds a gingko seed.

SECONDARY CELL WALL: Any layer of cell wall (cellulose and other polysaccharide mesh) that has been glued with pectin onto the top of a pre-existing primary cell wall to add layers of structural support. There is a trade-off between added protection and homeostatic ability, as each extra cell wall adds a level of difficulty in transporting materials back-and-forth. Plasmodesmata are critical to the exchange of materials between cells with extra walls.

SECONDARY GROWTH: Refers to lateral growth that widens the girth of a plant. There are several centers of meristematic tissue that divide to do this. The vascular cambium adds layers of phloem on the outside and xylem on the inside, contributing rings to a tree during each growth season. Cork cambium divides to produce layers of cork that thicken the bark. Lateral meristems throughout a plant divide to produce side branches and roots.

SECONDARY MERISTEM: Any meristematic tissue, such as cork cambium and vascular cambium, that is resposible for dividing and differentiating to new tissues that increase the girth or lateral spread of the plant.

SECONDARY MESSENGERS: Intracellular signal molecules of various types that are released or activated in response to an environmental cue (such as in phytochromes) or the triggering of a cell receptor (such as plant hormonal responses). Typically, the secondary messengers lead to a pathway that activates transcription factors that turn on genes and regulate the rate and amount of protein production.

SECONDARY PHLOEM: Found in the trunks and woody stems of gymosperms and perennial dicot angiosperms, it is produced on the lateral exterior surface of the vascular cambium as it divides. Primary phloem is found in all plant tissues, but in the trunk or stem, it is located on the outer periphery of secondary phloem in rays.

SECONDARY ROOTS: Refers to roots that grow from lateral meristems to allow roots to spread laterally in the soil. The size, number, and orientation depend upon the type of plant. Gymnosperms and dicot angiosperms tend to have smaller secondary roots that emerge from a large taproot, while monocot angiosperms with fibrous roots often have many similarly-sized lateral secondary roots.

SECONDARY XYLEM: The new layer of yearly xylem that develops from dividing meristematic tissue on the interior surface of the vascular cambium. It forms a new ring of growth each year, has many layers of cell walls, and is primarily responsible for the transport of water and soluble nutrients from the roots to the trunk and leaves.

SEED: The fertilized ovule of a gymnosperm or angiopserm plant. A seed includes an embryo, a food supply (endosperm and/or cotyledons), and a protective seed coat. It connects to the ovary (fruit) via the pedicle to allow it to be transported away from the parent by animals or wind. Seeds represent a massive survival improvement over spores, allowing gymnosperms (and later angiosperms) to largely replace bryophtes and pterophytes in forests.

SEED COAT: A sclereid-rich fibrous covering that protects a developing embryo and its food source from dessication and consumption. Splits open upon germination to allow the embryo to develop.

SHORT DAY PLANT: A type of plant that is triggered to flower by phytochromes exposed to longer hours of darkness with the approaching Fall equinox or lengthening days as Winter turns to Spring. Will not bloom with too many hours of Summer daylight or an excessive night lengths of Winter. Examples of these include Christmas cactus, chrysanthemum, and poinsettias.

SIEVE TUBE ELEMENT: A living donut-shaped type of phloem cell that is responsible for the transport of sugars and soluble nutrients through the phloem. There are large numbers of membrane pump proteins on the surface of sieve tubes. Most of the cytoplasm and organelles have been removed from sieve tubes and the nucleus is absent, allowing it to focus on a single task. Helpless to maintain itself, it must be cared for by companion cells.

SILIQUE: A pod-shaped dry dehiscent fruit that resembles legumes, but only splits open along one natural seam to release the seeds. Brassicas, such as mustard and turnips produce siliques, as do wallflowers and radishes.

SIMPLE LEAF: A leaf that grows independently of others as a single blade and petiole. Very common. Examples include banana trees, oaks, maples, comfrey, cherries, and thousands of others.

SEPAL: A modified leaf that forms a concentric whorl underneath the petals. In most flowers, sepals close and protect immature flower buds. Usually green, but many monocots (especially lilies) have colored sepals that are used to attract pollinators in conjunction with the petals.

SEPTUM: A dry brittle membrane that forms discrete pockets around seeds inside of a silique (a dry dehiscent fruit that splits open on one seam). Examples of plants with septa around their seeds are shepherd's purse and mustard.

SPERM: The haploid male gamete that fertilizes haploid female ovules to create zygotes that develop into plant embryos. Generated from germ-line microspore mother cells inside the antheridium of bryophytes and pterophytes, the germ-line cells inside of male pollen cones, or in the anthers of flowering plants. Two sperm are released from each grain of pollen, with one fertilizing the egg and the other a diploid cell to form an endosperm.

SOLANALES: An order of angiosperm dicots that contains an assortment of fibrous herbaceous plants, woody shrubs, and vines. Commonly have lobed leaves covered with trichomes. Often produce toxic alkaloids to deter herbivores. Flowers have nectaries under stamens and fruits are usually berries. Includes tomatoes, potatoes, nightshades, eggplants, and tobacco.

SOLUTE POTENTIAL: Measured in liter bars per mole, it describes the amount of osmotic pull a solution has relatie to the other. The lower the solute potential, the more hypertonic it is relative to a solution of higher value. Describes the direction that water will flow and how much that osmotic potential can offset gravity. For instance, a sugar beet will have a lower water potential than water in the soil, so it will pull water in, regardless of gravity.

SPONGY PARENCHYMA: Found in the leaves of C3 plants, they are rounded ground tissue cells that have chloroplasts to produce food for the plant via photosynthesis. As opposed to pallisade mesophyll, they are fatter, rounder, and scattered, with air spaces between them to allow CO_2 and O_2 exchange.

SPORE: The asexual reproductive cells generated by sporophyte plants in that stage of alternation of generations. Spores have half the original number of chromosomes as the parent plant, are generated by meiosis, and develop into either a male or female gametophyte plant.

SPOROPHYTE: The asexual diploid or polyploid stage of an organism that uses the alternation of generations life cycle like plants. Sporophytes are the larger spore-producing diploid stage of bryophytes, the larger leafier stage of ferns and other pterophytes that produce spores in sori, and are the green leafy 'typical' parts of the plant in gymnosperms and angiosperms, as the gametophytes are microscopic and hidden inside flowers.

SPRING WOOD: In woody dicot angiosperms, these are colloquially known as the light rings in a cross-section of wood. In Spring, water is typically more abundant and nutrients from stored food are flowing through the tree, so the vascular cambium divides more rapidly and produces larger cells. Collectively, the larger xylem has a lighter appearance than the more tightly packed dark rings of Summer wood.

STAMEN: The male reproductive organs of angiosperms. Comprised of a filament and an anther, the former lifts the latter up to the level of the female pistil for pollination. The anther contains the microspore mother cells that divide meiotically to generate pollen granules that enclose the sperm.

STAMINATE CONE: The male reproductive organs of gymnosperms. In monoecious cone plants, they are found only on male plants, while in hermaphroditic conifers, they are located on branch tips independent of female seed cones. Male cones are usually brush-like, with the microspore mother cells that generate pollen protected by a spire of modified leaves that form a soft spire.

STELE: Refers to all the collective vascular tissue at the center of a root. Sealed in by the endodermis in roots, the xylem forms an 'X' shape at the center of the dicot stele, with phloem sitting under its arms. In monocots, a concentric ring of scattered phloem surrounds a circle of xylem at the center. In stems, the

STEM: One of the 3 main organs of the sporophyte, its main functions are transduction of sugars and nutrients and transport of water between the roots and leaves, structural support, and elevating leaves to maximize the amount of light they receive for photosynthesis. Many plants have modified stems with other functions, such as the food storage in the underground rhizomes of ginger.

STIGMA: The sticky pad at the top of the female flower pistil that traps pollen grains for fertilization of the ova down at the base of the pistil in the ovary.

STIPULES: Small leaf-like appendages at the base of the petiole of typical leaves that are found in many angiosperms. In addition to secondary photosynthetic functions, they protect tender new leaves from damage.

STROBILUS: The cone-like reproductive structures that emerge from the shoot tips of the sporophyte generation in certain pterophyte plants, such as club mosses and horsetails. Also another name for the cone structures that house the gametophytes of gymnosperms.

STROMA: The location of the Calvin cycle in chloroplasts. The stroma is a soup of photosynthetic enzymes and biochemical intermediates. Enzymes in the matrix receive ATP and NADPH from the light reactions of the grana, use the energy and hydrogen atoms to assemble carbohydrates, and return spent ADP and NADP+ to the grana.

SUBSIDIARY CELLS: Cells adjacent to the guard cells that exert osmotic control over them by pumping water and ions back and forth, resulting in turgid guard cells opening the stomata, or flaccid guard cells closing them.

SUMMER WOOD: In woody dicot angiosperms, these are colloquially known as the darker thinner rings in a cross-section of wood. In Summer, water is typically less abundant and nutrients from stored food have largely been used, so the vascular cambium divides more slowly and produces smaller cells. Collectively, the smaller xylem has a darker appearance than the loosely packed rings of Spring wood.

SUPERASTERID CLADE: A large clade of angiosperm dicots that are grouped together phylogenetically because of DNA and RNA sequence similarities. As compared to superrosid ovules, the nucellus (embryo sac) is much thicker. There are 20 different orders in this clade, including more primitive groups that contain sandalwoods, azaleas, cacti, oleanders, as well as the mainline asterids that include asters, zinnias, hollies, nightshades, and carrots.

SUPERROSID CLADE: A large clade of angiosperm dicots that are grouped together phylogenetically because of DNA and RNA sequence similarities. As compared to superrosid ovules, the nucellus (embryo sac) is much thinner. There are 18 orders in the clade, including those that contain alders, grapes, willows, legumes, elms, figs, roses, apples, beeches, oaks, geraniums, squashes, and myrtles.

STYLE: A female flower organ that forms a tube that connects the pollen-trapping stigma to the ovary at the base of the pistil. It provides a pathway for a pollen tube to grow into ovules inside the ovary, during fertilization.

STOMATA: Openings found on the undersurface of leaves (with the exception of aquatic plants) that allow the uptake of CO_2 for the Calvin cycle of photosynthesis or for C4 uptake by PEP carboxylase. When the guard cells adjacent to them are turgid, they swell outward and open the pore. When they are flaccid, they sag and close the stomata to prevent water loss.

SUCCULENCE: An evolutionary adaptation seen in a number of plants native to arid, hot, and/or tropical regions to conserve water and avoid dessication. Succulent leaves are fleshy and swollen in order to store large volumes of water. Examples include cacti, euphorbias, aloes, stonecrops, and agaves.

SYMBIOSIS: A close permanent evolutionary relationship between two species. Can be mutualistic, with benefits for both, or commensalistic, with one species gaining an advantage, while the other is neither harmed nor helped.

SYMPORT: A type of membrane transport protein that uses a favorable concentration gradient difference in one substance to generate osmotic force that drags another desired substance into the cell against its concentration gradient. For instance, certain plant cells force sucrose in against its concentration gradient by allowing a flow of proteins through the same pump. There are binding sites for both substances.

SYNERGID CELLS: A pair of haploid cells that are generated from meiotic divisions of megaspore mother cells in the ovary, alongside the ova. The synergids flank the ova and send biochemical signals to guide the pollen tube to the ova.

TAPETUM: A layer of cells inside the anther of angiosperm plants that nourish and care for developing pollen.

TAPROOT: A long central prominent root found in gymnosperms and many angiosperm dicots that penetrates deeply into the soil vertically from an apical meristem at the root cap. Many taproots exhibit secondary growth and can be woody or impregnated with starch and other stored food molecules in their cortex. Other secondary meristems in the root generate smaller lateral roots.

TENDRIL: Found in vining plants, it is a leaf modified for climbing a surface, such as a tree trunk, for support and as a means of gaining access to light. Often grow in a spiral pattern around the support as a result of thigmotropism.

TERMINAL BUD: The apical meristem at the top of a shoot tip that generates the topmost leaves and branches. The center of primary growtAAh at the top of the plant.

TETRADYNAMOUS STAMENS: Found among plants of the Order Brassica, these appear in flowers as two elevated central pairs of stamens (four total) rising above the level of two exterio and inferior stamens.

TETRALOCULAR OVARY: An ovary that uses the ovarian wall to form four distinct divided lobes with separate seed compartments. A common type of arrangement in members of Order Brassicaceae.

THALLUS: A sheet-like leaf structure that makes up the main body of liverwort bryophytes. Rhizoids on the undersurface anchor it to a substrate. Lacks vascular tissue and resembles the leaf-like thallus of green macroalgae.

THIGMOTROPISM: Tendency of a plant to grow disproportionately on the side of the stem opposite a substrate that it touches. Allows vines to climb and wrap themselves around support structures and allows trees to grow around obstacles. Occurs because signal molecules trigger greater production of auxins and other hormones on one side of the plant, triggering disproportionate cell division and lengthening.

THYLAKOID: A disc-like membranous sac found inside chloroplasts that is the site of the light reactions of photosynthesis. Impregnated with chlorophyll and other pigments, their membranes have embedded electron transport chains that generate an electrochemical gradient that regenerates ATP and NADPH from photon energy.

TRACHEID: A wedge-shaped water-conducting xylem cell that utilizes pinholes in the cell wall to allow water to flow laterally through the xylem by capillary action. Unlike more evolved vessel elements, they lack tend-to-end channels between cell walls that form microscopic water pipes in the wood fiber. Most numerous and most common in gymnosperms, though dicot angiosperms have some tracheids, especially basal dicots.

TRANSCRIPTION: The process that converts the code of an activated gene in the DNA of the nucleus into copies of messenger RNA that are used by ribosomes to manufacture the prescribed protein needed. In the process, the DNA opens and RNA polymerase enzyme synthesizes a complementary mRNA from the sense strand of the DNA.

TRANSCRIPTION FACTORS: Intracellular proteins that control the rate and amount of DNA transcription into mRNA. They may be activated by secondary messengers or feedback loops and interact as part of a cascade of such proteins. They typically bind upstream of the promoter sequence of DNA and interact with RNA polymerase.

TRANSLOCATION: The movement of sugars and other substances through the vascular tissues of a plant. Photosynthetic products are made in the leaves and pumped downward to the stems and roots. Translocation may reverse in deciduous plants in early Spring when they need energy to produce leaves for the season ahead.

TRANSPIRATION: A process that is partially responsible for the upward movement of water through the stems or trunks of trees. Cohesion of water molecules and adhesion of the water to the cell walls of the xylem forms a column that moves upward by capillary action, but also because of negative pressure when the leaf stomata are open. Because there is greater water pressure inside the tree, suction pulls the water upward.

TRICARPELLARY FLOWER: A flower that has three ovaries within each corolla. The number of ovaries is independent of the number of locules in each. A common arrangement seen in many monocots, such as lilies.

TRICHOMES: Small hair-like projections that emerge from the epidermis in certain types of leaves. Trichomes may be used to trap moisture from dew or fog, such as in lamb's ear, or they may sharp and contain glands at their base that generate irritating substances to discourage herbivores, as in thistles or okra.

TRILOCULAR OVARY: An ovary that uses the ovarian wall to form three distinct divided lobes with separate seed compartments. A common type of arrangement in many monocots, such as onions, garlic, and other alliums.

TROPISM: A tendency of a plant to show differential growth towards a stimulus. Controlled by phytochromes and plant hormones, such as auxins, they are the result of greater cell division in one region of the plant. Tropisms include heliotropism (sun), phototropism (light), thigmotropism (touch), chemotropism (nutrients), and gravitropism (upward away from gravity in stems, toward gravity in roots).

TUBE CELL: A type of cell found in pollen grains that evolved in gymnosperm plants and was retained in flowering angiosperms. The tube cell eliminates the need for sperm to use an external water source, such as creek water or rain, to swim to the ova. Instead, it forms a moisture-rich pathway for the sperm to swim down the stigma of the flower to the ovule at the base of the pistil.

TUBER: A modified underground stem, such as in potatoes, impregnated with starch and nutrients. Allows long-term dormancy in cooler climates until warmer weather returns and meristematic tissue regenerates new leaves.

TURPENE: Aromatic compounds found in conifers and certain other plants that are derived from isoprene and sterols. Particularly common in conifers, but also found in many flowering plants, their purposes include serving as repellants to grazing animals or as attractants for pollinators.

UNDIFFERENTIATED TISSUE: In plants, it contains cells that are not yet fated to for their specialization as dermal, vascular, or ground tissue. Since none of their genes have been switched on or off, their maturation and development depends on various stimuli and plant needs. The equivalent of stem cells in animals, in plants, undifferentiated tissues are found in meristems.

UNILOCULAR OVARY: A plant ovary that has no sub-compartments or internal divisions. All ovules develop inside a single large ovary. Berries, pepos, and legume pods are all unilocular ovaries.

VASCULAR TISSUE: One of the three main types of tissue found in the organs in a plant, it is responsible for the movement of water, nutrients, and sugars through the tissues of a plant. Generated by the vascular cambium, it includes the tracheids and vessel element cells of xylem (water pipes) and the sieve tube elements and companion cells of phloem (sugar translocation). OAAne of the major advances of plants over ancestral green algae.

VESSEL ELEMENT: One of the two types of cells found in xylem, they are shaped like fluted pipes with angular faces on their vertical surfaces. Openings in these faces allow water to travel rapidly upward by capillary action. Vessel elements are a major advance over tracheids, as they carry water much faster. The rato of vessel elements to tracheids is highest in angiosperm dicot hardwood trees. They are absent in gymnosperms other than gnetophytes.

VERNALIZATION: A period of cold dormancy during Winter that is necessary for the proper hormonal cues that trigger seed germination in many types of plants native to temperate and cold climates, such as the seeds of apples, peaches, cabbages, turnips, and daffodils.

VITALES: An order of angiosperm dicots that belong to the Superrosid clade. Most are woody vines with flowers that lack obvious petals. Instead, they form a structure called a calpytra around a single conical pistil surrounded by five stamens. Their fruits are usually berries. Includes grapes, muscadines, cissus, and Virginia creeper.

WATER POTENTIAL: A measurement expressed in liter bars per mole that predicts the direction of water movement between two solutes of differing concentrations. Solutions with relative lower water potential are hypertonic to solutions with higher water potential, resulting in a net movement of water toward them.

WAXES: A class of lipids that are composed of long chain fatty acids bonded end-to-end around an ester group. Almost biochemically inert and unable to be utilized for energy by most life, they make an excellent sealant that keeps water from evaporating out of leaves and herbaceous stems

WOODY GROWTH: Secondary growth originating in the vascular cambium of gymnosperms and angiosperm dicot trees, shrubs, and woody vines. It is comprised of rings of secondary xylem layered inside of a newly generated circle of primary xylem. Woody growth typically does not emerge in angiosperm trees until the third year, wherein the cork cambium also generates periderm tissues that replace dermal tissue on the trunk.

XANTHOPHYLL: A yellow accessory pigment found in chloroplasts and chromoplasts that is derived from lipids. It absorbs wavelengths in the purple, blue, and green range, including some that chlorophyll misses. It is particularly abundant in yellow flowers (such as sunflowers) and yellow fruits (such as squash and bananas).

XYLEM: A type of vascular tissue used by plants to transport water. Functionally dead at maturity, its cells contain dozens of concentric cell walls that are impregnated with lignin and supported laterally by sclereid cells and other fibers. As they die, the middle becomes hollow, allowing capillary action to pull water up through the stem or trunk. Two types of xylem exist. Tracheids are wedge-shaped and less efficient than flute-shaped vessel elements.

VASCULAR BUNDLE: A cluster of conducting vessels found throughout the leaves and stems of a plant. Typically, phloem vessels are on the outside of the bundle and xylem are the interior. They are surrounded by various fibers composed of sclerenchyma and collenchyma in herbaceous stems and by bundle sheath cells in leaves.

VASCULAR CAMBIUM: A ring of living secondary meristamatic tissue that is found between the phloem and xylem of a stem or trunk. It divides laterally on both sides, differentiating into phloem on the outside and xylem on the inside. It is the tissue that grows and increases the girth of the stem or trunk.

VEGETATIVE NUCLEUS: One of the four nuclei generated meiotically by microspore mother cells. It ends up with the lion's share of nutrients and cytoplasm because it is fated to become the tube cell that grows into the pistil as a conduit for sperm to the ovule.

VEGETATIVE REPRODUCTION: One of the two forms of asexual reproduction that plants use to form clones of themselves. Can refer to generation of plantlets, pups, rhizomes, or other clones by budding or fragmentation.

VERSATILE ANTHERS: Common among grasses, such as bamboo and corn, this type of anther has a single point of attachment between the stamens and the filament in the middle of the dorsal surface.

ZINGINERALES: An order of angiosperm monocots that are often found as understory plants in tropical forests. Leaves tend to have very large blades with thick midribs and grow from a sheath-like apical meristem. Production of aromatic compounds and succulent stems are common. Flowers are spike-like with whorled inflorescence. Many are C4 plants. Includes ginger, bananas, galangal, turmeric, prayer plant, an bird-of-paradise.

ZYGOTE: A fertilized diploid (or polyploid) ovum that is fated to divide meiotically and differentiate into the tissues as the embryo develops.

CAKE TRIVIA ANSWERS from page 55

1) Pound cake traditionally calls for one pound each of butter, sugar, flour, and eggs. Sour cream pound cakes also use a pound of sour cream. They're the original healthy dessert.
2) Billy the Kid was offended that he was offered chocolate cake. 'Ah ah ah! Pendleton! The white cake with the SWEET frost!'
3) Marie Antoinette never said 'Let them eat cake!' The phrase actually came from a book written by Jean Jacques Rousseau. If you want to know the context, read it yourself. I just can't trust supposed 'geniuses' who wore powdered wigs and knickers.
4) Applesauce can be substituted for oil and mashed bananas can be substituted for eggs. However, do it in the wrong type of cake and it will taste off. Everyone will know you didn't want to bother going to the store and they will gossip about you.
5) Cherry Liquer. Also anger. The entire German language sounds angry. Play Rammstein and scream at the cake as it bakes.

MIGHT A BRIGHTER FUTURE AWAIT? CONCERNED CITIZENS SAY 'NO MORE!' TO THE BARBARISM OF UNCONTROLLED PRODUCE CONSUMPTION.

www.ingramcontent.com/pod-product-compliance
Lightning Source LLC
Chambersburg PA
CBHW081212130726
47997CB00009B/2632